# MANAGING GLOBAL GENETIC RESOURCES

## The U.S. National Plant Germplasm System

# MANAGING GLOBAL GENETIC RESOURCES

## The U.S. National Plant Germplasm System

Committee on Managing Global Genetic Resources:
Agricultural Imperatives

Board on Agriculture
National Research Council

NATIONAL ACADEMY PRESS
Washington, D.C.   1991

NATIONAL ACADEMY PRESS • 2101 Constitution Avenue, NW • Washington, DC 20418

NOTICE: The project that is the subject of this report was approved by the Governing Board of the National Research Council, whose members are drawn from the councils of the National Academy of Sciences, the National Academy of Engineering, and the Institute of Medicine. The members of the committee responsible for the report were chosen for their special competences and with regard for appropriate balance.

This report has been reviewed by a group other than the authors according to procedures approved by a Report Review Committee consisting of members of the National Academy of Sciences, the National Academy of Engineering, and the Institute of Medicine.

This material is based on work supported by the U.S. Department of Agriculture, Agricultural Research Service, under Agreement No. 59-32U4-6-75. Additional funding was provided by Calgene, Inc.; Educational Foundation of America; the Kellogg Endowment Fund of the National Academy of Sciences and the Institute of Medicine; Monsanto Company; Pioneer Hi-Bred International, Inc.; Rockefeller Foundation; U.S. Agency for International Development; U.S. Forest Service; W. K. Kellogg Foundation; World Bank; and the Basic Science Fund of the National Academy of Sciences, the contributors to which include the Atlantic Richfield Foundation, AT&T Bell Laboratories, BP America, Inc., Dow Chemical Company, E.I. duPont de Nemours & Company, IBM Corporation, Merck & Co., Inc., Monsanto Company, and Shell Oil Company Foundation.

**Library of Congress Cataloging-in-Publication Data**

The U.S. National Plant Germplasm System / Committee on Managing
   Global Genetic Resources: Agricultural Imperatives.
      p.    cm.—(Managing global genetic resources)
   Includes bibliographical references and index.
   ISBN 0-309-04390-5
   1. National Plant Germplasm System (U.S.)  2. Crops—United
States—Germplasm resources.  3. Germplasm resources, Plant—United
States—Management.  I. National Research Council (U.S.).
Committee on Managing Global Genetic Resources: Agricultural
Imperatives.  II. Title: US National Plant Germplasm System.
III. Title: United States National Plant Germplasm System.
IV. Series.
SB123.3.U17  1991
631.5'23'0973—dc20                       90-21056
                                            CIP

Any opinions, findings, conclusions, or recommendations expressed in this publication are those of the author(s) and do not necessarily reflect the view of the organizations or agencies that provided support for this project.

Printed in the United States of America

# Committee on Managing Global Genetic Resources: Agricultural Imperatives

PETER R. DAY, *Chairman,* Rutgers University
ROBERT W. ALLARD, University of California, Davis
PAULO DE T. ALVIM, Comissão Executiva do Plano da Lavoura Cacaueira, Brasil*
JOHN H. BARTON, Stanford University
FREDERICK H. BUTTEL, Cornell University
TE-TZU CHANG, International Rice Research Institute, The Philippines
ROBERT E. EVENSON, Yale University
HENRY A. FITZHUGH, International Livestock Center for Africa, Ethiopia†
MAJOR M. GOODMAN, North Carolina State University
JAAP J. HARDON, Center for Genetic Resources, The Netherlands
DONALD R. MARSHALL, Waite Agricultural Research Institute, Australia
SETIJATI SASTRAPRADJA, National Center for Biotechnology, Indonesia
CHARLES SMITH, University of Guelph, Canada
JOHN A. SPENCE, University of the West Indies, Trinidad and Tobago

## Genetic Resources Staff

JOHN A. PINO, *Project Director*
MICHAEL S. STRAUSS, *Associate Project Director*
BRENDA E. BALLACHEY, *Staff Officer*
JOSEPH J. GAGNIER, *Senior Project Assistant*

---

* Executive Commission of the Program for Strengthening Cacao Production, Brazil.
† Winrock International, through January 1990.

# Subcommittee on Plant Genetic Resources

**ROBERT W. ALLARD,** *Chairman,* University of California, Davis
**AMRAM ASHRI,** The Hebrew University of Jerusalem, Israel
**VIRGIL A. JOHNSON,** University of Nebraska
**RAJENDRA S. PARODA,** Indian Council of Agricultural Research,
New Delhi
**H. GARRISON WILKES,** University of Massachusetts, Boston
**LYNDSEY A. WITHERS,** International Board for Plant Genetic Resources,
Italy

## National Plant Germplasm System Work Group

**CALVIN O. QUALSET,** *Chairman,* Genetic Resources Conservation Program,
University of California, Davis
**JOHN L. CREECH,** U.S. Department of Agriculture (Retired)
**S. M. (SAM) DIETZ,** U.S. Department of Agriculture (Retired)
**MAJOR M. GOODMAN,** Department of Crop Science, North Carolina State
University
**A. BRUCE MAUNDER,** DEKALB Plant Genetics, Lubbock, Texas
**DAVID H. TIMOTHY,** Department of Crop Science, North Carolina State
University

# Board on Agriculture

The National Academy of Sciences is a private, nonprofit, self-perpetuating society of distinguished scholars engaged in scientific and engineering research, dedicated to the furtherance of science and technology and to their use for the general welfare. Upon the authority of the charter granted to it by the Congress in 1863, the Academy has a mandate that requires it to advise the federal government on scientific and technical matters. Dr. Frank Press is president of the National Academy of Sciences.

The National Academy of Engineering was established in 1964, under the charter of the National Academy of Sciences, as a parallel organization of outstanding engineers. It is autonomous in its administration and in the selection of its members, sharing with the National Academy of Sciences the responsibility for advising the federal government. The National Academy of Engineering also sponsors engineering programs aimed at meeting national needs, encourages education and research, and recognizes the superior achievements of engineers. Dr. Robert M. White is president of the National Academy of Engineering.

The Institute of Medicine was established in 1970 by the National Academy of Sciences to secure the services of eminent members of appropriate professions in the examination of policy matters pertaining to the health of the public. The Institute acts under the responsibility given to the National Academy of Sciences by its congressional charter to be an adviser to the federal government and, upon its own initiative, to identify issues of medical care, research, and education. Dr. Samuel O. Thier is president of the Institute of Medicine.

The National Research Council was organized by the National Academy of Sciences in 1916 to associate the broad community of science and technology with the Academy's purposes of furthering knowledge and advising the federal government. Functioning in accordance with general policies determined by the Academy, the Council has become the principal operating agency of both the National Academy of Sciences and the National Academy of Engineering in providing services to the government, the public, and the scientific and engineering communities. The Council is administered jointly by both Academies and the Institute of Medicine. Dr. Frank Press and Dr. Robert M. White are chairman and vice-chairman, respectively, of the National Research Council.

# Preface

The work of sustaining a productive agriculture by developing new crops and improving existing ones is inextricably linked to the need for plant germplasm—the broad array of materials, from seeds to large trees, that are maintained and preserved because of the genetic traits they may contain. Germplasm manipulation led to the production of the high-yielding varieties of rice, wheat, and maize that now play a key role in world food production. The conservation and exchange of germplasm have become important concerns for many nations and international institutions. This report examines the management of plant germplasm in the United States and the activities of the National Plant Germplasm System (NPGS). It may also serve as a springboard for discussions about the establishment of a U.S. program that encompasses a broader array of genetic resources.

The NPGS holds a wide array of germplasm collections and is the world's most active distributor of germplasm samples to other nations. The scientific, technical, and support staff at many of its facilities are dedicated to preserving these important resources for future generations. The committee recognizes the past important contributions made by many dedicated individuals who have contributed to the achievements of the NPGS.

As the size and number of NPGS collections have grown over the years, the task of managing them at a number of sites located throughout the United States has also increased. As a result, the national system has become a loose association of sites and organizations that, at times, has a diversity of goals and lacks coordinated leadership. The NPGS will continue to grow and it must continue to respond to emerging needs. It must now become a distinct, centrally managed, and nationally coordinated unit of the U.S. Department of Agriculture. This report is directed toward accomplishing that end.

The Committee on Managing Global Genetic Resources, established by the National Research Council in November 1986, is concerned with the world's genetic resources that have an identified economic value. These resources are important to agriculture, forestry, fisheries, and industry. The committee has been assisted by two subcommittees and several work groups that gathered information or prepared specific reports. One of the work groups, chaired by Calvin O. Qualset, examined the U.S. National Plant Germplasm System and drafted this report. It is one of five committee reports to be published in a series entitled, *Managing Global Genetic Resources*. Other reports prepared by the committee, its subcommittees, and work groups address issues related to the global management of forest trees, livestock, fish and shellfish, and crop plants. The examination of crop plants will be included in the committee's main report, which will address the legal, political, economic, and social issues surrounding global genetic resources management as they relate to agricultural imperatives.

In addition, a work group was appointed to provide information that would aid in planning and designing a new storage facility for the National Seed Storage Laboratory. The committee released a letter report, *Expansion of the U.S. National Seed Storage Laboratory: Program and Design Consideration*, in April 1988. Copies of this report are available from the Board on Agriculture.

The charge to the NPGS Work Group was to assess the structure and operation of the components of the national system; to determine its adequacy in applying technology as well as anticipating and meeting the present and foreseeable needs of U.S. agriculture; and to examine the policies guiding its interaction in the international conservation and use of plant genetic resources. Within the context of the committee's work, examination of this system provides a valuable case study for similar programs in other countries. More specifically, the work group was asked to do the following:

- Assess the extent to which the organization, administration, and operation of the present system meet the needs of exploration, collection, conservation, and use of agricultural plant germplasm.
- Consider to what extent present arrangements encourage or constrain cohesiveness and coordination, and promote wide participation within the system.

- Determine the appropriate roles for the various elements of the system at national and state levels, as well as the nature of the interaction between the public and private sectors in germplasm conservation.
- In the context of meeting the needs of U.S. agriculture, define the possible interactions of the NPGS with other national and international programs, and examine the potentials for international cooperation for collection, conservation, or distribution and the adequacy of U.S. policy in this regard.
- Define priority needs of the national system, in terms of funding, existing technologies, facilities, personnel, and research.
- Define and assess the nature of linkages between base collections, working collections, and in situ collections as these relate to the work of the NPGS and plant breeders.

The report begins by outlining the challenges of managing plant germplasm, describing the components of a national system, and reviewing the origins of germplasm management in the United States and the development of the NPGS. Chapter 2 details the sites, collections, laboratories, and offices that comprise the national system. Chapter 3 describes the administrative components of the NPGS and how the system obtains and acts upon advice. The committee's recommendations make up the last chapter. The committee offers these to aid the NPGS in continuing and enhancing its impressive record of service to U.S. and global agriculture.

Appreciation of the importance of plant germplasm and its impact on the development of modern agriculture can easily be obscured amid the details of multiple administrative relationships and site descriptions. An understanding of how a small, weedy plant growing in a distant land can be important to this process is often lost. For this reason the report contains summaries of six individual germplasm samples and the diverse ways in which they came to be recognized and used.

PETER R. DAY, *Chairman*
Committee on Managing
Global Genetic Resources:
Agricultural Imperatives

# Acknowledgments

Many scientists and administrators of the U.S. National Plant Germplasm System have contributed time, support, and information essential to the development of this report. The committee acknowledges the assistance of individuals at the regional plant introduction stations, the national clonal repositories, and the National Germplasm Resources Laboratory (formerly the Germplasm Services Laboratory), and thanks the many advisory groups within the NPGS for allowing committee members and staff to attend meetings and participate in discussions.

Several individuals provided information used in developing the profiles of specific germplasm accessions. The contributions of David W. Altman, L. Gene Dalton, Donald N. Duvick, Kenneth J. Frey, Edgar E. Hartwig, Larry J. Grauke, Frederick R. Miller, Charles M. Rick, Zea Sonnabend, Tommy E. Thompson, Noel Vietmeyer, Roger D. Way, Kent Whealy, Denesse Willey, Thomas Willey, and Johnny C. Wynne are gratefully acknowledged. The committee also appreciates the personal perspectives on germplasm work of Raymond L. Clark and Stephen Kresovich.

Administrative and secretarial support during various stages of the development of this report was provided by Philomina Mammen, Carole Spalding, and Maryann Tully, and they are gratefully acknowledged.

Special thanks is extended to Henry L. Shands, national program leader for plant germplasm, who contributed his knowledge and time to answering the many inquiries for information about U.S. germplasm activities.

# Contents

# The U.S. National Plant Germplasm System

# Executive Summary

Agricultural production in the United States has grown remarkably. One of the greatest periods of growth occurred between 1930 and 1980, when U.S. yields of corn, wheat, and potato increased 333 percent, 136 percent, and nearly 300 percent, respectively. Roughly half of these increases in crop yields are attributable to genetic improvements, which also led to varieties with better nutritive value and greater pest, disease, and stress resistance. The genes necessary for this crop improvement are contained in a broad array of plant materials, which when used in breeding or genetic research are termed germplasm.

Increased agricultural production has contributed significantly to the U.S. economy. Agricultural exports accounted for $28 billion, or 12 percent of total domestic exports, in 1987, the most recent year for which complete data are available. Crops and food products accounted for 81 percent of these exports. Cash receipts from the sale of crops in the United States totaled $72.6 billion in 1988, an increase of $10.7 billion from the previous year. Plants also have significant economic value to pharmaceutical, fiber, chemical, and other industries.

Sustaining agricultural productivity will require continued use of and access to a broad diversity of germplasm. Managing genetic resources, therefore, is a strategic necessity for the United States.

Preservation of the tissues, seeds, and plants that comprise the nation's plant germplasm resources is the responsibility of the National Plant Germplasm System (NPGS), a diffuse network of laboratories and research stations. The NPGS is a federal and state cooperative effort.

1

Its activities are supported at the federal level by the Agricultural Research Service (ARS) and the Cooperative State Research Service (CSRS) of the U.S. Department of Agriculture (USDA), and at the state level by state agricultural experiment stations. Since the early 1970s, the primary responsibility for management and support has rested with the ARS.

The size and scope of many NPGS collections and the volume of national and international distributions of samples from them are noteworthy. Indeed, many NPGS collections are considered to be valuable and important global resources. This accomplishment is testimony to the tireless efforts of many scientists, technicians, and other support staff.

As agricultural scientists and plant breeders improve crops, their need for germplasm will grow; the NPGS must keep pace. At the same time, concerns about the loss of biological resources place an ever greater significance on germplasm management and conservation, and on a growing international role for the NPGS. By conserving the genetic diversity of crop species and their wild relatives, the NPGS contributes to national and international efforts that address the loss of biological diversity.

To meet these increasing demands, the NPGS must be a centrally managed organization. At present, it exists within a decentralized framework in which a multitude of individuals, committees, and USDA offices have varying levels of responsibility. This framework has hampered the ability of the NPGS to function as a coordinated, well-defined system with clear-cut leadership, responsibilities, and authority. It also constrains the resolution of long-standing needs and problems.

To meet both national and global needs, the NPGS must recognize and act on the needs of the nation's germplasm collections. It must be guided by budgetary procedures that invest resources in areas of need and opportunity, and it must utilize evaluation and planning mechanisms that identify systemwide needs and cost-effective solutions to recognized deficiencies. The committee's basic conclusion is that it will remain very difficult, if not impossible, for the system to function properly without a major overhaul in its structure and administrative procedures.

This report presents recommendations that, when implemented, will further strengthen the NPGS as a viable and effective conservator of the nation's genetic resources. It examines the administration, management, and activities of a national system that traces its beginning to 1898, yet operates without a national structure. The committee concludes that the growth and ever-increasing national and international importance of plant germplasm mandates the adoption of a new approach to systemwide management, and it offers two options for achieving this goal. It considers the more viable option to be the establishment of the NPGS as an independent agency within USDA's Office of Science and Education.

## GERMPLASM: A RESOURCE AND A RESPONSIBILITY

Germplasm includes older and current crop varieties, specialized breeding lines used to develop new varieties and hybrids, landraces of crops that have emerged over centuries of selection by farmers, wild plants related to individual crops, and mutant genetic stocks maintained for research, particularly when gathered together in organized collections of plants, seeds, or tissues. Germplasm collections can range from plants maintained in greenhouse or field plantings, to dried seeds in sealed envelopes held at low temperatures, to in vitro cultures of tissues or buds.

The phenomenal agricultural productivity of the United States has come from using germplasm to improve crops genetically. Examples include pest and disease resistance in tomatoes from obscure, wild *Lycopersicon* species related to the cultivated tomato and the derivation of modern corn and wheat from early landrace varieties introduced by Mexican Indians and European and Middle Eastern settlers, respectively. Modern high-yielding wheats are derived in part from semi-dwarf varieties introduced from Japan following World War II. Much of the germplasm in U.S. collections originated in other countries.

The viability of U.S. agriculture depends on a flow of enhanced crop varieties that can withstand pests, diseases, or climate extremes. The adaptability of pests and diseases, changes in agricultural practice, such as greater emphasis on the biological control of pests, and changes in consumer needs or preferences require the continuation of the varietal development process.

Genetic diversity is a natural resource. All nations share the responsibility for its management and the privilege of using it. Individuals, private organizations, and universities can contribute to the maintenance of germplasm, but the tasks of overseeing and managing genetic resources are clearly beyond the capacity of any individual or group. Coordination of the many and varied efforts in the United States, including international collaboration, must be through a national, government-supported, centrally managed program.

Germplasm activities in the United States have been largely driven by an unofficial policy of national self-sufficiency that calls for comprehensive collections to reduce dependence on other nations or institutions. However, global cooperative efforts to manage biological resources are of increasing importance for managing collections housed in the United States and for arranging the most appropriate and economical replenishment of seed supplies.

The nation's participation in and support of international cooperation in managing germplasm will continue to grow. The NPGS is the world's largest distributor of plant germplasm. Each year it supplies more than 230,000 samples from its collections to more than 100 nations. Eighteen specific U.S. crop collections, including those of maize, rice, sorghum, wheat, soybean, citrus, tomato, and cotton, have been designated by the International Board for Plant Genetic Resources (IBPGR) as regional or global base collections in its international network. The United States also provides back-up storage to other collections, such as that of the International Rice Research Institute in the Consultative Group on International Agricultural Research system.

Economic issues, such as trade balances, property rights, and international cooperation in agriculture and the conservation of biological diversity, make it imperative for the United States to provide support and leadership in defining and implementing domestic and international programs. The NPGS must foster international cooperation to protect the world's biological resources and preserve public and private sector access to genetic diversity for the benefit of all nations, many of which exhibit unique environmental or agroclimatic conditions.

## AN OVERVIEW OF THE SYSTEM

The management of plant germplasm was formalized in 1898 when efforts to introduce useful plants were concentrated in the newly created USDA Plant Introduction Office. The Agricultural Marketing Act of 1946 led to the creation of the USDA regional plant introduction stations in the late 1940s and early 1950s, and to the opening of the National Seed Storage Laboratory (NSSL) in 1958. These facilities were established to conserve germplasm, to foster its use in plant breeding, and, for the NSSL, to provide secure, long-term storage. A national program began to emerge. In the early 1970s, the NPGS arose as a collaborative federal and state attempt, with some cooperation from private industry, to better manage the germplasm of importance to U.S. agriculture.

Although often regarded as a well-defined entity, the NPGS is constrained by the absence of a clear delineation of its duties, programs, and sites, nor does the NPGS budget process lend itself to systematic management and timely initiative in areas of critical need or opportunity.

The association of such elements as the regional stations, NSSL, and the National Small Grains Collection with the NPGS is readily apparent, but many others are loosely associated. Even more important is that the system has no central administrative control, and hence it has limited capacity to identify or act on needed changes in program activities and germplasm management methods. Other assessments of the system have termed it a diffuse network (Council for Agricultural Science and Technology, 1984; Office of Technology Assessment, 1987). For example, the NPGS has no central office and staff, nor a definitive location in any USDA organizational structure. Laboratories and scientists, particularly within the ARS, may be considered to be part of the NPGS by virtue of their work on managing or enhancing germplasm.

### Collections and Facilities

The national system's collections contain more than 380,000 different accessions of some 8,700 species, including virtually all of the crops of interest to U.S. agriculture. The collections are managed at various laboratories and facilities located around the country. Sites or units of major importance are the

- National Seed Storage Laboratory located in Fort Collins, Colorado, for long-term, back-up storage of the NPGS collections. Of its more than 230,000 accessions, about 60,000 are not duplicated at other sites.
- Four regional stations in Pullman, Washington; Ames, Iowa; Geneva, New York; and Griffin, Georgia. They are responsible for the management, regeneration, characterization, evaluation, and distribution of the seeds of more than one-third of the accessions of the national system (i.e., nearly 135,000 accessions of almost 4,000 plant species).
- National clonal germplasm repositories at 10 locations in the United States, including Puerto Rico, for conserving and managing fruit, nut, and other species that cannot be held in seed collections (more than 27,000 accessions of nearly 3,000 species).
- National Small Grains Collection in Aberdeen, Idaho, which is responsible for more than 110,000 accessions of wheat, barley, oats, rice, rye, *Aegilops* (a wild species related to wheat), and triticale (a hybrid of wheat and rye).
- Interregional Research Project-1 (IR-1) in Sturgeon Bay, Wisconsin, which holds about 3,500 potato germplasm accessions, including cultivated forms of the white or Irish potato and more than 100 related wild species.
- Several crop-specific collections in universities or USDA labora-

tories that are devoted to maintaining and managing particular species. Among these are the cotton collection in College Station, Texas (more than 5,500 accessions); the long-season soybean collection in Stoneville, Mississippi (more than 3,700 accessions); and the short-season soybean collection in Urbana, Illinois (nearly 10,000 accessions).

The NPGS is intended to address the needs of its primary users— breeders and other researchers. It is important to structure and document NPGS collections so they can locate the accessions most likely to possess the genetic traits sought. While national collections may only occasionally be used by breeders, the traits (i.e., genes) extracted from them are often passed among breeders and other researchers and can benefit many breeding programs.

Various USDA offices are responsible for data management, acquisition, and quarantine. The Germplasm Resources Information Network (GRIN), intended to be a comprehensive database of NPGS holdings, is managed through the National Germplasm Resources Laboratory (NGRL), formerly the Germplasm Services Laboratory, in Beltsville, Maryland. The Plant Introduction Office and activities related to planning and coordinating plant exploration and collection are also part of the NGRL. The National Plant Germplasm Quarantine Center, also part of this laboratory, facilitates the movement of imported plant germplasm through quarantine and into NPGS collections in cooperation with the USDA Animal and Plant Health Inspection Service.

## Support for the System

It is difficult to determine precisely the actual costs of the national system. The ARS provided about $26.5 million in fiscal year 1988 for germplasm-related activities. About half of this amount was used to fund the sites and collections of the national system. Support for the principal NPGS sites and collections totaled about $13.8 million; the remaining $12.7 million went largely to evaluation and enhancement activities conducted mostly at other ARS sites. In addition, CSRS provided about $900,000 in 1988 to the four regional stations and $132,000 to IR-1. Facilities, equipment, services, and personnel at NPGS collections are frequently provided as in-kind support by the state agricultural experiment stations and universities where germplasm facilities are located. Private industry has also provided support for specific projects or activities, such as the Latin American Maize Project, which seeks to evaluate a broad range of maize germplasm from Latin America. Although cooperation has brought together diverse scientific interests and expertise, it has complicated NPGS management and administration. For example, several different, independent advisory groups and administrative lines oversee virtually every major site.

## LEADERSHIP AND ADVISORY FUNCTIONS

Leadership and advisory functions within the national system are difficult to discern. The evolution of the system has produced numerous committees and individuals with varying degrees of authority and responsibility. As the lead agency for NPGS management, the ARS administers its programs through a decentralized system of area offices and its National Program Staff. This approach impedes the efficient, coordinated management of what should be a nationally oriented system.

The NPGS has no clearly designated, central administrative leadership. As a member of the ARS National Program Staff, the national program leader for plant germplasm is charged by the ARS with planning responsibility for the national system. He or she, however, has little authority over budgets, programs, or management at individual sites, and can only offer management recommendations. These recommendations require approval from the Germplasm Matrix Team (composed of other National Program Staff with different research planning responsibilities) and the concurrence of the deputy administrator of the ARS, the administrator, and then the relevant area directors, each with multiple and different priorities. Thus, the program leader cannot ensure the implementation of programs conceived as part of planning responsibilities.

The CSRS also provides regional research funds to the NPGS as mandated by the Hatch Act (Public Law 84-352). The national program leader for plant germplasm has no authority over the distribution or use of these funds.

Individual sites, such as a regional plant introduction station, are responsible independently to each of their funding authorities, which may include ARS, CSRS, and a state agricultural experiment station. This creates parallel and duplicate sets of authorities, responsibilities, policies, and procedures for many sites.

Providing advice for managing the national system is no less complex. Many committees and individuals hold varying and frequently overlapping advisory responsibilities. These include the

- National Plant Genetic Resources Board (NPGRB), which according to its charter advises the secretary of agriculture and the National Association of State Universities and Land-Grant Colleges on national policies and priorities relating to the acquisition, management, exchange, and use of plant germplasm. Scientists and administrators from the public and private sectors (including universities) are appointed for 2-year terms, and may be reappointed twice.

• National Plant Germplasm Committee (NPGC), the intended function of which is to guide and coordinate the system by developing policies, priorities, and proposals related to funding, research, and international relations. Members are drawn from the ARS, CSRS, experiment stations, and the private sector.

• Crop advisory committees that provide expert advice on acquisition, management, and use for particular crops or crop groups (e.g., wheat, beans, leafy vegetables, or woody landscape plants). There are currently 39 of these committees comprised primarily of scientists with crop-related expertise.

• Technical committees, established by CSRS for the NPGS sites, that are funded by CSRS or through an experiment station. They consist of representatives from each of the state agricultural experiment stations in the region and from appropriate federal agencies.

• Technical advisory committees that provide advice to individual national clonal repositories and are composed of scientists with technical expertise in one or more of the crops maintained at that site.

• Plant Germplasm Operations Committee, assembled by the ARS national program leader for plant germplasm for the purpose of discussing specific questions or actions and operations between and within the national system's sites. It is an ad hoc assembly of ARS research leaders and site managers of major NPGS sites or activities, but it has no direct authority over the NPGS.

• Germplasm Matrix Team, chaired by the national program leader for plant germplasm and comprised of the ARS agricultural science adviser for plant germplasm and the ARS national program leaders responsible for research planning on commodities or subjects generally related to germplasm use (e.g., range, pasture and forage crops, plant health). The at times competing areas of responsibilities of these individuals must be balanced against concerns about plant germplasm.

## FINDINGS AND RECOMMENDATIONS

The NPGS, as presently constituted, has no discernible structure and organization. It lacks a central, clearly defined authority and process for managing its activities, formulating national policies, identifying priorities, or developing budgets necessary to act on new policies and emerging priorities. NPGS is managed by too many individuals, committees, and USDA offices. The USDA can remedy these defects by creating a more centrally managed system. It must take systematic actions in six critical areas: administration (especially in linking the budget process to key system needs), germplasm acquisition and collections, facilities and personnel, the mission of the national system, data management, and research. This section presents an overview of the committee's recommendations. A more detailed discussion with additional recommendations appears in Chapter 4.

## Administration

*The administrative and advisory organization of the National Plant Germplasm System should be structured to provide for efficient national coordination.*

The need to coordinate nationally a variety of activities and agencies and to respond to growing international relationships has made efficient management of the national system an imperative. The system's management structure must be made more compatible with its nature and activities. For example, the conservation, management, and distribution of germplasm are service activities. At present, the NPGS is largely supported and managed by the ARS, a research agency that functions through regional area offices. More direct authority and responsibility for budget and programs of the NPGS must be vested in a centralized management unit to enable the system to respond more effectively to national needs and priorities.

Effective national coordination of the NPGS depends on establishing a management structure that links programs and policies to budgetary authority and budget process outcomes. The authority to formulate budget recommendations in accordance with the identified needs and responsibilities of U.S. germplasm efforts must reside with an office or individual intimately associated with the operation of the NPGS. This approach will also reduce the complexity of NPGS decision making and funding processes. Through a coordinated, national structure the NPGS could also take the scientific and technical lead in guiding U.S. germplasm activities with other nations and in the international community.

Funds already designated by USDA offices for acquisition, preservation, and evaluation activities could make up the budget for this unit. The portion of the $26.5 million in ARS funds for these activities in fiscal year 1988 was $21 million ($22.5 million in 1989), of which $13.8 million supported work at the principal NPGS sites. The CSRS provided an additional $900,000 in 1988 for management costs at some sites. More funding could be needed for selected enhancement activities, to accommodate the growing size of collections, to develop and operate desert and subtropical sites, and to regenerate accessions. All of these areas, however, represent current needs of the NPGS, and they are not related to administrative restructuring.

## Options for Achieving National Coordination

To achieve more centralized national management will require administrative and structural changes to the system and the way it is organized within the U.S. Department of Agriculture. Two options are proposed.

*Organization Outside the Agricultural Research Service*  The national system could be established as an independent entity within USDA's Office of Science and Education. It would be responsible for all aspects of the program, including budget formulation, staff, operations, and site management. Funding would encompass the budgets already designated for germplasm activities by ARS and CSRS. This approach would enable the execution of a national germplasm program through a central authority. Furthermore, the leader of the NPGS would report directly to the assistant secretary for science and education, and could be granted authority comparable to the heads of other USDA research agencies.

This new structure would provide the national system with direct lines of leadership and authority, and would enable the NPGS to respond directly to specific needs in establishing priorities, programs, and budgets. It would unambiguously establish the NPGS as an organization whose leaders have greater visibility and control of budgets, policies, and operations and more direct line authority over sites and individuals. Program and budget guidance would be provided by the NPGRB, which represents both the participating agencies and offices and the user community.

However, this option has potential disadvantages. Outside ARS, the NPGS would stand alone as a relatively small program, competing for budgets and political support against three much larger, well-established USDA science and education agencies, the ARS, the CSRS, and the Cooperative Extension Service. It may become difficult for the NPGS to obtain cooperation from larger services, and its visibility in the USDA budget process might lessen. Separation from ARS may also distance germplasm work from the basic research that has been important to advancing NPGS activities. Finally, the ARS provides administrative support (e.g., services related to personnel, contracts, accounting, and purchasing) that would have to be created for a new unit, and which could entail additional costs.

*Elevation Within the Agricultural Research Service* Under this second option, the national system would remain in the ARS, but it would be strengthened by adoption of a range of changes in management policies, budget process, and facilities. Leadership would be vested in an individual or office directly answerable to the ARS administrator. The influence of the area directors would be reduced and germplasm program planning would not be the responsibility of the National Program Staff or the Germplasm Matrix Team.

The budget for the NPGS should be separate and distinct from other ARS activities and should be directly related to identified needs and priorities. An annual budget for the national system would be developed by its leader. This budget would address recommendations, concerns, and priorities identified by the NPGRB. It should account for all ARS funds devoted to NPGS activities and include consideration of funding supplied cooperatively by other agencies.

It is clear that significant change is needed in the administration of the NPGS and in the mechanisms for formulating budgets to make them responsive to identified needs. Addressing the committee's recommendation solely through cooperative or informal agreements that perpetuate the present lines of authority over the NPGS budget and program will do little to address the committee's concerns or the system's needs. The committee considers the option to establish the NPGS as an independent unit outside the ARS to be the most likely to achieve efficient national coordination. It is aware, however, that other arrangements may be more readily achieved within political and federal budget constraints. The second option of remaining within ARS but in an elevated status would confirm the leadership role of the ARS in managing and financing the NPGS, but it could lead to reduced willingness of other agencies to provide cooperative support.

### Advisory Groups

Under a centrally managed NPGS, advisory groups should each have clearly defined responsibilities, the resources to accomplish their assigned tasks, and an established pathway for assuring that their advice is used to formulate policies and procedures. These groups include the NPGRB and crop advisory committees.

*The National Plant Genetic Resources Board must have greater independence as an adviser on national and international policies.*

The NPGRB is chaired by and reports to the assistant secretary for science and education, a design of its charter that can constrain its ability to address controversial issues. To make it a more dynamic, responsive, and independent adviser on U.S. plant genetic resources activities, the board should elect a chair from its membership, and its members should equally represent the government offices or agencies contributing to U.S. germplasm activities and the broad user community, including industry, universities, and the private nonprofit sectors.

The board should provide budgetary and program guidance to the leader of the NPGS and make recommendations on the policies, priorities, and activities that comprise U.S. germplasm efforts. It should prepare for the secretary of agriculture, the Congress, and others an annual report on U.S. germplasm activities and the effectiveness of the NPGS in achieving the board's budget and program recommendations.

*The National Plant Germplasm Committee should be disbanded.*

When originally established in 1974, the committee was a source of information about, and an advocate for, the national system. Over the years, committee membership has become more of an administrative obligation, which has led to the appointment of several representatives who lack direct responsibilities for, or involvement in, the NPGS. Today there is no clear role for the NPGC that is distinct from other advisory groups.

*The crop advisory committees should be provided financial support, and a mechanism should be created to use their reports when developing policies and priorities.*

Crop advisory committees prepare reports on national and international developments concerning specific crop species. They discuss implications for the United States and make recommendations for strengthening NPGS activities. Although a potentially important source of knowledgeable and technical information and advice for the national system, the committees receive no financial support and there is no established mechanism for using their reports. Some financial support for travel and administrative expenses would enable those members without other resources to participate in meetings. The information contained in reports could be gathered and analyzed, and used to aid the national system in establishing its priorities and resource allocations. The responsibility for analyzing reports could be assigned to an existing committee or to one drawn from the committees' chairs. Furthermore, liaison between chairs and the NPGRB needs to be established so their substantive policy concerns are addressed.

### Germplasm Acquisition and Collections

*Collections must be managed as national, not regional, resources.*

The distinction between regional plant introduction stations, which are viewed as being supported by both the ARS and the states within their respective regions, and national clonal germplasm repositories, which are considered as having more national focus, should be eliminated. These facilities should be designated as national plant germplasm centers. Elimination of the administrative differences would promote cooperation among all of the centers and would simplify the system's structure.

*Curators with specific knowledge should be appointed for each major crop or crop group, and they should be given management responsibilities.*

There is now no plan to ensure that knowledgeable, suitably trained curators oversee acquisition and management of the major or essential collections in the national system. At present some site managers oversee several crops. Curators must have specific knowledge about their crop plants and be familiar with their collection, documentation, regeneration, evaluation, and enhancement. They should work with the appropriate crop advisory committee and the leader of the NPGS to develop and implement plans for the management and enhancement of germplasm.

*The National Plant Germplasm System must devote more of its resources to regenerating seed accessions.*

Regeneration of seed lots with low germination is a continuing need. A large proportion (almost 50 percent) of the accessions at NSSL are below the minimum desired size (550 seeds). Regeneration of these samples is urgently needed. Where responsibility for providing fresh seed cannot be assigned to an existing site, funds should be available to secure regeneration on a contract basis with appropriate supervision and safeguards. Where regeneration is required outside the United States the NPGS should make contractual arrangements with the appropriate international groups or foreign government agencies.

*A plan should be developed for monitoring, supporting, and conserving important special collections.*

Special collections have proved to be invaluable, both to the scientific understanding of genetic processes controlling plant growth, development, and physiology and to the development of improved crop varieties. They include collections developed by individual researchers for specific crops and those assembled for basic research. Administration of special crop collections in ARS is complicated because funding and program responsibilities related to them are intermingled with other commodity-based programs. For important special collections, back-up storage should be provided within the NPGS.

*The management of large collections, such as those for wheat, corn, and soybeans, could be aided by the identification of core subsets, but this method must be applied cautiously.*

The identification of core subsets of no more than 10 percent of the total accessions has been considered as a potential management tool for large collections. Subsets of collections could be useful for setting priorities to evaluate specific characteristics, such as disease resistance. Breeders who find core samples with the genetic traits they seek could use them as guides to select others in larger collections from, for example, the same regions or environments that possess germplasm with similar traits. However, a misunderstanding of the nature and purpose of core subsets could result in irreparable damage to collections. Core subsets are not a means of eliminating redundancy or of combining or bulking accessions. Further, while these subsets should receive priority for evaluation, the maintenance of other accessions must not be neglected. Core subsets are inappropriate for collections that are small or that represent a limited geographic area.

## Facilities and Personnel

*The National Seed Storage Laboratory must be expanded.*

The present structure of the NSSL, described in the report, *Expansion of the National Seed Storage Laboratory: Program and Design Considerations* (National Research Council, 1988), is antiquated and insufficient to meet the needs of the national system. Further, the capacity does not exist for the laboratory to provide full back-up storage for the NPGS collections. Appropriation of adequate funds to expand the facility in accordance with its growing needs and responsibilities should be a national priority.

*Facilities and programs of the National Plant Germplasm System should undergo periodic external review.*

National germplasm centers and crop-specific collections should be reviewed regularly to ensure that their programs, resources, and staff capabilities meet the needs of the germplasm for which they are conservators. Reviewers should have scientific or management experience in plant germplasm or its use, and should be drawn from outside the national system. The appropriateness of a site's location should also be considered.

While it is not advisable for collections to be moved frequently, the location of collections should be based on scientific considerations and opportunities for cooperation with universities or experiment stations. It is, however, essential that collections be situated in areas that are ideal for their growth and performance. Where support from an experiment station has declined, the potential to move the activity to another location should not be overlooked.

*Sites should be established for the growth and maintenance of germplasm that requires short day-lengths or arid environments.*

Many collections of corn, cotton, and beans include accessions from tropical areas that flower and produce seed only during short days. For example, some accessions in the bean collection at the Regional Plant Introduction Station in Pullman, Washington, must be grown during winter in a heated greenhouse, making them expensive to maintain. Constraints, such as the need to keep materials disease free or to prevent cross-pollination, may necessitate the use of such controlled environments. For many accessions, however, a suitable facility with short day-lengths and cool but frost-free conditions must be found in a tropical environment. Arrangements could be made with another nation or an appropriate international organization. An irrigated site in the arid desert of the southwest United States where diseases and pests are less abundant is also needed to maintain the germplasm of crops, such as jojoba, some sorghum and beans, several grasses, and selected small grains.

### The Mission of the National System

*The National Plant Germplasm System should develop clear, concise goals and policies that encompass the conservation of plant genetic resources that reflect the world's biological diversity and crop resources of immediate use to scientists and breeders.*

Assessments of the scope and extent of current collections are needed, and these must be used to develop long-range management plans. Efforts are needed to expand some collections to make them more representative of the available diversity. Consideration must be given to ecogeographical areas from where accessions originate, how broad based or narrow based the collection is in terms of known or suspected genetic traits, and what genes might be obtained by utilizing various transect or other sampling procedures when rare alleles are sought. These factors must be weighed against cost, accuracy, need, and other criteria for obtaining suitable material from recognized collection centers in other areas of the world.

Collection and conservation priorities should be established to address collection completeness and include close wild relatives and non-crop-related species that may possess useful genes. A policy should encompass crop genetic resources and endangered species of native and exotic taxa, and should include consideration of in situ conservation, where appropriate.

*The United States must address the problem of global loss of biological diversity. This can be done in significant part through conserving the genetic diversity of crop species.*

The increasing responsibilities of the United States to contribute to and expand efforts for preserving global germplasm resources must be recognized. In this context, the NPGS must be a proactive system that develops long-term plans and policies for broader collections that encompass a greater range of diversity. This commitment must be reflected in the mission of the national system, and it must be supported by the USDA, the Department of State, and the Congress.

### International Policies and Cooperation

*The National Plant Germplasm System must take a more active role in developing U.S. policies that guide relations with the Food and Agriculture Organization, international agricultural research centers, and other international agencies and national institutions.*

While the NPGS has accepted responsibility for several collections that are designated as part of an international network, the Department of State defines and manages international relations. No cohesive, scientifically based policy exists to guide the nation's international activities related to plant genetic resources.

Policies for international cooperation in germplasm activities should permit an unambiguous mechanism for establishing U.S. positions on international genetic resources issues. In the past, it has been unclear where responsibility for recommendations on U.S. positions resides. Thus, concerns expressed by scientists, administrators, or advisers with regard to germplasm policy seemed to disappear. The National Plant

Genetic Resources Board, as an adviser on germplasm issues, should provide a forum for the discussion of these issues.

*The United States should become a member of the Commission on Plant Genetic Resources of the Food and Agriculture Organization of the United Nations.*

Joining the commission should not be considered an endorsement of all of its actions or policies. Rather, it would enable the United States to gain a voice in developing and directing the commission as well as shaping its long-term agenda. The Department of State should include the NPGS and the NPGRB in the process of developing policies and actions for consideration by the commission. The leader of the national system or his or her designee should be part of any U.S. delegation to commission meetings. [The United States joined in September 1990.]

*The National Plant Germplasm System should cooperate with other nations to conserve, collect, maintain, and regenerate germplasm.*

Many nations are now becoming reluctant to allow the indiscriminate collection and exchange of germplasm. The United States must seek policies that promote open and cooperative collection, management, and exchange, and that include opportunities to promote in situ conservation of important resources. Cooperation would benefit and enlarge the quantity of germplasm and technical expertise available to other nations.

The United States should pursue agreements with other national or international germplasm centers for access to or regeneration of important germplasm resources. The NPGS holds, for example, accessions of Andean Maize landraces, but there are no U.S. facilities suitable for regenerating these high-elevation, short day-length materials. Cooperative agreements are needed with other nations to regenerate such germplasm.

*The United States should work with neighboring countries to establish a North American cooperative program in genetic resources.*

Canada and Mexico have national germplasm systems, both of which are smaller than the NPGS, but may be complementary to it. In a regional cooperative program, the United States, Canada, and Mexico would mutually benefit. The United States, for example, could obtain assistance in maintaining or regenerating accessions for which no suitable environment is available domestically. Many cotton accessions, for example, must be grown under contract at sites in Mexico. Duplication of the U.S. wheat collection with Agriculture Canada to provide a supplementary backup to that in the NSSL is another example of cooperation. Similarly, these neighbors can benefit from cooperation with NPGS facilities, which can provide, for example, back-up storage, testing, or technical assistance.

## Information Management

*The Germplasm Resources Information Network must better reflect the collections of the National Plant Germplasm System.*

The GRIN is an effective database management system for the NPGS collections, but its inventory and descriptive data are incomplete. For many accessions there is little specific information beyond the accession number, crop name, and where it is held. Data on geographic origin (passport data) and basic characteristics (characterization data) may exist within NPGS records, but they often are not available through the network. These kinds of data are essential to breeders and other researchers who seek particular genetic characteristics. Much greater emphasis must be placed on making such data available through GRIN.

## Research

*A research advisory committee should be established to assess and guide the system's research activities.*

Research is essential to sustain an effective genetic resources management system. For example, seed biology studies are the basis for developing nondestructive methods for assessing seed viability; cryobiology research could lead to extended storage of in vitro cultures or short-lived seed; population genetics efforts could improve collection, maintenance, and regeneration methods; and biotechnology research may enable more efficient and rapid use of the genes contained in germplasm collections. No mechanism for adequately reviewing or promoting research presently exists. An advisory committee could assess research needs, resources, and accomplishments and could consider research proposals from scientists in the NPGS. It could also oversee periodic external peer review of NPGS research. Existing in-house peer review lacks scientific and technical rigor.

*Funds should be made available for competitive, goal-directed research in areas of specific need.*

Competitive grants should be used to fund research for which expertise is not available within the national system. The program could be managed either by the NPGS or within the existing USDA Competitive

Research Grants Office. A competitive grants program would enable the national system to promote goal-directed research in areas of specific need for which permanent staffing is inappropriate. Framing guidelines for proposals and awarding grants could be the responsibility of the research advisory committee, proposed above.

With the appropriate administrative organization, the NPGS would become an effective mechanism for ensuring agricultural security, both nationally and internationally. The dedication of NPGS scientists and the significance of their work have long been recognized. The impact of their contributions, however, has often been hampered by an inadequate administrative structure. The changes proposed in this report are intended to remedy this situation.

# 1

# Managing Crop Genetic Resources

Plant germplasm activities in the United States have evolved since the colonists first brought seeds from their home countries and exchanged them with native Americans. Introductions of new crops and crop varieties to the United States were vigorously pursued from the late nineteenth century through the first decades of the twentieth century. Over the years, the reasons for assembling collections of crop germplasm have become more compelling. They now include concern over loss of biological diversity as well as the economic importance of accessibility to the germplasm resources needed to sustain national food supplies.

Germplasm resources are a strategic resource essential to national and global agricultural security. Recent technological advances such as cell and tissue culture, cryopreservation, and recombinant DNA (deoxyribonucleic acid) technologies provide the potential for innovations in preserving and using plant germplasm. As new technologies are developed it will be important for the United States to adopt those that can enhance its genetic resource conservation capabilities. The United States must also adopt policies and procedures to assure adequate preservation of resources, commensurate with its expanding activities and international responsibilities.

## GERMPLASM AND GERMPLASM COLLECTIONS

Germplasm is the term used to describe the seeds, plants, or plant parts useful in crop breeding, research, and conservation efforts. Plants,

21

seed, or cultures are germplasm when maintained for the purposes of studying, managing, or using the genetic information they possess. Thus, seed of an old, heirloom tomato variety is just seed when produced by a gardener or seed company, but it is germplasm when part of a collection gathered to conserve the genetic diversity of tomatoes or to develop a breeding program for new tomato varieties, or even for the purpose of preserving particular genetically controlled traits.

Today's scientists and crop breeders must have access to, and knowledge of, a wide array of crop varieties, landraces, and related wild species in their search for specific genetic traits. The seeds, pollen, or other plant materials in which these traits are found are called genetic resources, or germplasm. In the United States such materials are held in a cooperative network of federal and state facilities known as the National Plant Germplasm System (NPGS), the subject of this report.

A collection of germplasm usually includes primitive landraces and wild species related to particular crops, and developed varieties and breeders' lines. Germplasm is used to develop new plant varieties for food, feed, fiber, turf, forages, and ornamentals and for forestry, industrial, and medicinal purposes. Hence, germplasm—the seeds and plants used as building blocks in breeding new cultivars—may be similar to the plants grown by a farmer or gardener, or they may be quite different and even further removed from the produce purchased in a market.

Some germplasm needs little or no breeding to be useful. Many commonly used ornamental, medicinal, and herbal plants have been changed little from their wild progenitors. Plants, such as the popular ornamental, Bradford pear, were originally brought to the United States as germplasm (Creech and Reitz, 1971). By contrast, some germplasm accessions may be of value only for the individual genetic traits they possess. Residents of New England and the Northern Plains who grow yellow roses, particularly varieties that bloom early, can do so because the plants contain genes for hardiness from the *Rosa xanthina*, which was introduced early in the twentieth century by the famous plant explorer Frank N. Meyer (Cunningham, 1984).

Germplasm collections include many kinds of materials used in crop breeding and development (Brown, 1989a,b; Chang, 1985; Creech and Reitz, 1971; Ford-Lloyd and Jackson, 1986; National Research Council, 1972; Plucknett et al., 1987). A large portion of the germplasm of many crops includes current and obsolete varieties. Breeders' collections may consist of inbred lines extracted from hybrids or varieties, superior varieties, elite lines with special combinations of traits, and intermated populations of elite and selected material. These sources of germplasm are reasonably productive and are used most frequently by plant breeders. Other common sources are lines or varieties from elsewhere that may be unadapted to local growing conditions, markets, or industrial processes, but which may possess valuable characters. Obsolete varieties are also found in the collections of individuals and organizations that preserve the seeds of "heirloom" varieties.

Primitive landraces, indigenous varieties, and specifically adapted ecotypes are important genetic resources. Genetically heterogeneous, adapted to specific local environments, and often unsuitable in appearance or quality for modern markets, they can be rich sources of genes. The races of maize developed by the American Indian and the landraces of cereals, forages, vegetables, and ornamentals brought by immigrants formed the germplasm base from which U.S. agriculture developed. Landraces are the products of centuries of planting, selecting, and replanting by farmers. The areas of the world where they were developed contain considerable genetic diversity and are important collecting sites for germplasm.

Researchers have developed collections of genetic stocks for several

The Andean highlands, farmed for more than 2,000 years, are important sources of genetic diversity for the potato. Traditional landraces, still cultivated by present-day farmers, and naturally occurring *Solanum* species are found there. Credit: Calvin Sperling.

crop species. These accessions have unique mutant genes, groups of genes, or gene deletions or duplications; interchanged or translocated chromosomes, where two distinct chromosomes had broken and the parts of each were then interchanged; and chromosomes with portions that are inverted. Some of the stocks may have duplicate or deleted chromosomes or may represent genetically distinct cytoplasms.

Although not used by consumers, genetic stocks are essential to genetic and cytogenetic research and some plant breeding procedures. These include, for example, locating nonnuclear genes in chloroplasts or mitochondria and studying photosynthesis and cytoplasmic male sterility. For some crop species, many genes have been "mapped" according to their linear arrangement on a chromosome. These maps are extremely valuable for gene manipulation and basic research. Genetic stock collections are often difficult and expensive to maintain, and specialized knowledge, skills, and facilities are required to increase and maintain accessions.

Special germplasm collections accumulated through genetic and breeding research have been the basis for studies of speciation, evolution, or taxonomy; manipulations of genes, chromosomes, or entire genomes; and the incorporation of genes into crop plants from weedy relatives or related species. Some important special collections have become part of the NPGS. Many, assembled by researchers at state or privately supported universities and used by those researchers throughout their careers, are extremely vulnerable to loss and may be abandoned or discarded by later researchers with other interests.

## PRESERVING PLANT GERMPLASM

Seeds are the most commonly preserved form of germplasm. Those of the major cereal grains, legumes, and most vegetable crops can be dried and stored for long periods under low humidity and at low or subfreezing temperatures. After storage, such seeds germinate readily, although some may require specific conditions of temperature, moisture, light, or darkness. Sometimes, to overcome seed dormancy, environmental conditions such as light, temperature, and humidity must be manipulated. Even then, a few species or accessions may germinate poorly or slowly.

For several reasons germplasm of some crops may not be stored as seed. In these cases germplasm must be maintained as live plants in the field or under cover (e.g., greenhouse or screenhouse), pollen, tissue cultures, or cuttings (e.g., scion wood of fruit and nut trees).

Many important tropical crops, such as cocoa or *Hevea* rubber, and some temperate species, such as wild rice (*Zizania aquatica*), have seeds that are damaged when dried and cooled. These seeds have often been termed recalcitrant, but are more correctly described as desiccation sensitive. These species must be maintained as living plants in field plantings or in a greenhouse. Plants can be raised from seed or when a particular clonal genotype from vegetative propagules (e.g., cuttings, bud grafts) is required (e.g., for many tree fruits and nuts or sugarcane). Clonal maintenance is also necessary for crops that lack true seed or have seed that is rare or difficult to obtain (e.g., banana and garlic).

The emerging techniques of biotechnology have opened new opportunities for germplasm storage and use (Peacock, 1989). Tissue culture is increasingly used to preserve virus-free plants and to supplement field plantings of germplasm when seed storage is impractical. Embryos may be excised from developing seeds and grown on artificial media when, for various reasons, they would not be viable if left to mature. Cryopreservation—storage in or suspended above liquid nitrogen at temperatures from $-150°C$ to $-196°C$—may greatly extend storage periods. Finally, it may be possible one day to maintain isolated DNA routinely and to use it for crop improvement.

DNA contains the molecular sequences that comprise genes. Thus, an organism's isolated DNA contains all of its genes. The potential to store isolated DNA underscores the central importance of germplasm as genetic information. This technology is already becoming available. However, its general application to germplasm management and crop improvement will require considerably greater knowledge of gene structure, regulation, placement, and function than currently exists as well as more precise understanding of the genetic control process of important crop traits.

Larger human populations, widespread use of new high-yielding varieties, and agricultural development projects focus concern on the collection, maintenance, and evaluation of genetic resources that are disappearing from farmers' fields and their native environments. Society's increased awareness of its obligation to prevent loss of the world's biological diversity has served to broaden the scope of germplasm conservation to include a wider range of species. Ultimately, the size and scope of germplasm activities in the United States must enlarge and additional efforts will be needed to accommodate this expansion.

## THE CHALLENGES OF CONSERVING AND MANAGING PLANT GERMPLASM

The first challenge to a plant germplasm program is to acquire representative sets of samples of those species that merit conservation. Obtaining a genetically diverse collection usually means adequate sampling from the distribution range of the selected species. Crops already are represented by a broad range of diversity and samples may be obtained from the institutions holding them. For many species, however, collections are incomplete and require further exploration and expansion.

Since germplasm collections are established for long-term (in perpetuity) conservation, they should reflect deliberate decisions concerning the materials critical to the welfare of a nation's agriculture and environment, and the provision of long-term technical support and funding for maintenance. A well-defined and accepted policy should

*Wild rice can be found growing naturally in Minnesota. Credit: Calvin Sperling.*

## WILD RICE
### Zizania aquatica L.

GRIN Data

No wild rice accessions are listed in the GRIN.

Wild rice (*Zizania aquatica*) was a staple of American Indian life for hundreds of years, but it has been barely 20 years since the first successful commercial efforts were made to grow this reed-like aquatic grass in the United States. Native to the central and upper Great Lakes region and to New England, wild rice anchors itself with shallow roots in mud with its stalks extending upward through as much as 6 feet (2 meters) of water. Seventeenth century French explorers described the plant as *folle avoine*, wild oats, to which it bears closer resemblance than to cultivated rice.

It was not until the early twentieth century that wild rice appeared in markets. Favored for its taste and keeping qualities, wild rice was being served in restaurants in Minnesota's hotels. The concept of cultivating *Z. aquatica* seemed out of the question, so wild rice was collected for these markets from wild plants. The primary problem was that the seeds of the plant shattered, or fell off, upon ripening. The absence of shattering allows grains such as oats, wheat, and barley to be successful commercial crops. Early pioneering efforts to cultivate wild rice met with losses as high as 90 percent.

state the focus of the collections as well as their extent, goals, and mode of operation.

Managing a germplasm collection extends far beyond acquiring the plants or seeds that comprise it and holding them under conditions of long-term storage. The prime reason for assembling collections is to make them available for breeding and research. They are also insurance against loss of rare and endangered wild species that are conserved in situ. Regeneration, characterization, evaluation, and documentation are important parts of the activity of managing genetic resources that must be addressed by a program.

---

The breakthrough that allowed farmers to grow wild rice came somewhat unexpectedly from a routine inspection of a field in Waskish, Minnesota, in 1963. Dr. Paul Yagyu and graduate student Erwin Brooks noted that unlike most other plants in the field a few apparent mutants seemed to retain their male flowers long after shedding their pollen. While such a characteristic would not necessarily imply that the same plants would hold mature seed, the two scientists speculated that this might be the case. In fact, offspring of those plants did shatter less and led to development of shatter-resistant wild rice. Their genes are now in all kinds of commercial wild rice.

Today Minnesota and California grow wild rice commercially with about 26,000 acres under cultivation. The 7.6 million pounds of wild rice produced in 1988 was worth about $15 million. Canada, the only other country to produce wild rice commercially, accounted for about 2 million pounds in 1987. Wild rice has a long way to go before it can be considered truly domesticated. Although current crops are less shatter prone than most wild types, about half of the seed still falls before it can be collected. Additionally, plants are tall and produce a large amount of straw in relation to grain. Wild rice is vulnerable to brown spot, a serious fungal disease that is severe in the standing water culture system used for wild rice.

Like other crop improvement programs, researchers are seeking to locate additional wild rice plants and to discover useful traits that can be transferred to domesticated lines. It may be that there are additional traits hidden in wild *Z. aquatica* plants that will further improve this newly domesticated grain. But that effort is in its infancy, and there is no central facility that conserves wild rice seed for future breeding programs, in part because the seeds themselves must be stored at high-moisture content and have relatively short periods of viability.

---

*The narrative was prepared from information supplied by Noel Vietmeyer, National Research Council. GRIN = Germplasm Resources Information Network.*

Seeds do not live indefinitely. Regeneration or replenishment can be an expensive, labor-intensive, time-consuming, and important process that is all too often unappreciated. A range of sites and seasons with appropriate temperature, moisture, day-length, length of frost-free growing season, and appropriate cultural conditions must be available. Populations of plants large enough to minimize the loss of alleles and shifts of gene frequencies must be established. Few plants are needed for individual homozygous accessions of self-pollinating species, but several hundred may be necessary for maintaining variable populations of self- and cross-pollinating accessions to retain genetic diversity. Seeds of high quality can be stored for many years (e.g., 50 or more for most accessions of cereal crops or beans), reducing the frequency of regeneration and its accompanying risk of error, genetic drift, or destruction by pests, diseases, or natural calamities.

Among the most important information on each accession is the ecogeographical information and the specific plant and ecological data obtained at the time of collection. This is referred to as passport data. Each time a sample is grown for regeneration there is opportunity to collect data on plant height, flowering date, lodging, vigor and senescence, fruit color and size, relative yield, and other characters. Detailed information should be available for at least a representative set (sometimes called a core) of accessions of a species or crop. This will usually involve special screening tests designed to measure disease and insect resistance, tolerance to various field environments, replicated yield trials, or even yield evaluations of many stocks hybridized by an elite, adapted tester as the common other parent.

Germplasm that must be maintained as field or greenhouse collections can be lost in several ways. Pests and diseases can devastate germplasm when it is grown to replenish seed, or when maintained in field collections. Wind, hail, frost, and drought can affect the survival of field collections. The landrace collection of potatoes held by the Centro Internacional de la Papa (International Potato Center) in Lima, Peru, for example, has been damaged by weather in the past and would have been partially lost were it not for duplicates maintained at another site. The proximity of eastern filbert blight to the National Clonal Germplasm Repository in Corvallis, Oregon, may necessitate moving the collection to another more remote back-up site.

## STRUCTURE OF A GERMPLASM MANAGEMENT PROGRAM

The basic elements of germplasm management are shown in Figure 1-1. These are grouped into acquisition, conservation, management, and utilization. Acquisition includes developing priorities for adding to

```
┌─────────────────────────────────┐
│         ACQUISITION             │
│                                 │
│  Exchange                       │
│  Exploration and collection     │
│  Introduction                   │
│  Documentation                  │
│  Quarantine                     │
└─────────────────────────────────┘
                │
                ▼
┌─────────────────────────────────┐
│         CONSERVATION            │
│                                 │
│  Active/working collections     │
│  Permanent (base) collections   │
│  Back-up collections            │
│  Clonal collections             │
└─────────────────────────────────┘
                │
                ▼
┌─────────────────────────────────┐
│         MANAGEMENT              │
│                                 │
│  Multiplication  Characterization│
│  Regeneration    Evaluation     │
│  Testing         Documentation  │
│                  Distribution   │
└─────────────────────────────────┘
                │
                ▼
┌─────────────────────────────────┐
│         UTILIZATION             │
│                                 │
│  Breeding or enhancement        │
│  Research                       │
│  Biological diversity management│
└─────────────────────────────────┘
```

FIGURE 1 1 The basic elements and flow of materials within a germplasm management program.

the collections, exploring, and collecting. It also involves exchange with other collections or donations from various sources. Knowledge of existing collections and information about the genetic diversity of the crops in general help in setting exploration priorities. The aim of acquisition is to obtain the most genetically diverse collection possible of the crop and its useful relatives (Simmonds, 1979).

Accessions that can be stored as seed are conserved and maintained in active collections from which distribution to breeders and other users can be made. These must be backed up by secure base collections, held in long-term storage (for most seed this is at −18°C or below, and 5 to 7 percent seed moisture). Clonal germplasm and materials not easily

The U.S. collection of wheat germplasm is held in cold storage at the National Small Grains Collection in Aberdeen, Idaho. Credit: U.S. Department of Agriculture, Agricultural Research Service.

stored as seed are maintained in active field collections or in screenhouses or greenhouses. Tissue culture, in vitro storage, cryopreservation, and pollen storage provide alternatives for medium-term and long-term maintenance. Special sites or facilities are used for these purposes.

Management involves more than the storage or housing of stocks. Small seed lots may need to be multiplied to increase the sample size or to improve viability before they become part of a collection. Seed must be increased or regenerated when the viability of the stored sample declines or the sample becomes too small because of testing or distribution. Seed samples must be tested regularly to detect decline in seed viability. Collections must be characterized, evaluated, and documented to aid in selecting materials for breeding or research. Evaluations for particular traits of interest are performed either by users or by germplasm curators in response to immediate or anticipated needs. Core collections that represent the genetic diversity of the whole collection can aid researchers in identifying the potentially useful accessions in large collections of a single crop (Brown, 1989a,b). Documentation is the responsibility of curators.

The utilization of germplasm collections, until recently, was generally considered to be for the use of plant breeders and research scientists. However, collections, as reservoirs of biological diversity, including rare or endangered species, can aid in conservation and management efforts (Brown et al., 1989; Office of Technology Assessment, 1987). Ironically, breeders do not frequently seek new materials from germplasm collections (Duvick, 1984). In part this is because of the long time it may take to screen for desired traits and breed them into appropriate genetic backgrounds. Breeders depend heavily on data from evaluation trials to select materials for their breeding programs. Duvick found that breeders prefer to use the adapted materials produced by other breeders. Infrequent use also can result from inadequate information being obtained or disseminated about accessions. There are also cases where elite foreign germplasm is unavailable through the NPGS and the breeder must seek acquisition independently.

### The Diversity and Size of Collections

During the past 20 years, the maintenance of biological diversity has become an important part of national and international development activities. The realization that many species and primitive landraces face extinction has accelerated concern about and increased the potential size and scope of collections. An emphasis that includes a wider range of wild species related to crops may also increase the size of collections. Further, new technologies for gene transfer hold the promise that plant breeders may be able to move genes freely among distantly related species. The usefulness of collections thus would not be limited to the diversity within a species.

The degree to which a collection represents the genetic diversity available in the species is more important than its size. Nevertheless, capturing a wide range of diversity can require a relatively large number of accessions. Suggestions for ways of managing large, diverse germplasm collections to facilitate their efficient use despite their size have been proposed (Brown, 1989a; Chang, 1989). In the future, such practices as identifying a core or representative subset of NPGS collections could become increasingly important, especially to evaluate specific traits.

### Management and Global Responsibility

No single national plant germplasm system can assure the protection and conservation of all plant diversity or even all of the diversity among known economically important plants. Protection and conservation require the participation of many nations. A global strategy may eventually emerge from current discussions about conserving biological diversity. A global plan would consist of a mosaic of national and international collaborative efforts that together would assure effective conservation of biological diversity. In the interim, nations must do what they can to meet their respective needs.

In the United States, germplasm efforts must be organized as part of the emerging global system. Priorities and goals should reflect the responsibilities of participating in an international cooperative effort to manage plant genetic resources. The U.S. plant germplasm collection, one of the world's largest and most diverse, shares responsibility for preserving many unique landraces no longer obtainable in their countries of origin because of habitat loss and genetic erosion.

## PLANT GENETIC RESOURCES IN THE UNITED STATES

Most of the rich and varied plants of U.S. agriculture come from many regions of the world. Modern crops, each representing a broad array of genetic materials, reflect the transition from their wild origins through the process of domestication, farmer selection, and more recently intensive breeding.

When the European colonists first arrived in the New World they

---

### WILD OAT
PI 317757
*Avena sterilis* L.

GRIN Data

Cultivar: 6-1272-131
Origin: Israel
Maintenance site: National Small Grains Collection
Year PI assigned: 1966

An Iowa State University graduate student research project in the late 1970s used a wild-oat species that led to one of the most productive oat cultivars yet introduced in the United States. Plants of PI 317757, a wild species of oat collected from Israel over a decade earlier in the mid-1960s, were found to produce higher than average amounts of grain. However, like most wild grains, the seeds fell off the stalks (shattered) upon ripening, before they could be harvested.

The challenge was to transfer the high-yield factor to a cultivated variety while leaving the shattering tendency behind. Complicating this was the fact that some of the genetic traits desired from the wild plant were not found in the chromosomes of the nucleus, but were carried in smaller molecules of cytoplasmic DNA. Genes of this nature are inherited only

found an agriculture based on fruits, nuts, berries, beans, herbs, squashes, tobacco, and corn—all species unknown to them. The hard flint corns of the New England Indians contributed greatly to the development of modern maize varieties used around the world (Cox et al., 1988; Wilkes, 1988). Early settlers also brought many crops from their homelands to North America.

The introduction of new crops and crop varieties from other countries has continued in the United States since the early colonial days. This influx of plants, combined with distinct climates and soil types, has led to the development of a diverse agriculture. The gathering of plants from around the world was formalized near the end of the nineteenth century, and conserving and safeguarding them became a major concern in the 1940s (Purdue and Christenson, 1989; White et al., 1989).

## The Foundation of Crop Improvement

The growth in agricultural production in the United States has been remarkable. For example, yields of corn, wheat, and potato increased 333 percent, 136 percent, and nearly 300 percent, respectively, between 1930 and 1980 (Witt, 1985). In the economic arena, agricultural exports

---

from the female half of a cross. A breeding program had to be developed with this constraint in mind. The resulting hybrids contained both nuclear and cytoplasmic genes from the wild oat species.

By 1983, 20 separate lines had resulted from the breeding work and one, named Hamilton, proved more productive than the rest. It produced yields of 85 bushels per acre compared with about 70 bushels per acre for other varieties. But more significant, Hamilton is the first oat variety to possess cytoplasmic genes from a wild species. It is the interaction of these genes with those in the nucleus that have been found to enable Hamilton not only to produce well, but also to resist disease. The Hamilton cultivar was, for example, found to be resistant to barley yellow dwarf virus, a common disease of oats. It also has broad adaptability to different environments and is less susceptible to lodging, a condition in which the plant topples over from its own weight.

Other genes from PI 317757 have been used to raise the content of oil in oat seed. Such improvements may one day enable oats to become a profitable oil-seed crop, which would be especially useful at more northern latitudes where a limited number of crops can be grown as oil sources.

---

*"GRIN Data" for the plant introduction (PI) number above represent information contained in the Germplasm Resources Information Network (GRIN). The narrative was prepared from information supplied by Kenneth J. Frey, Department of Agronomy, Iowa State University.*

The viability of U.S. agricultural production depends on developing new and improved crop varieties. Credit: Pioneer Hi-Bred International, Inc.

accounted for $28 billion, or 12 percent of total domestic exports, in 1987 (U.S. Department of Agriculture, 1989a). Crops and food products accounted for 81 percent of these exports. Cash receipts from the sale of crops in the United States totaled $72.6 billion in 1988, an increase of 17 percent from the previous year (U.S. Department of Agriculture, 1989b). Plants also have significant economic value to pharmaceutical, cosmetic, and other industries.

Germplasm forms the foundation on which modern plant improvement rests. The genetic basis of high productivity in modern wheat, resistant to pests, diseases, and other stresses, was assembled by combining landraces and breeding lines from around the world. Plants used in breeding modern crop varieties often include related wild species. Most of the genes for pest and disease resistance in tomatoes come from wild *Lycopersicon* species. Promising experimental improvements in cotton fiber quality have come from apparently worthless wild *Gossypium* species having little or no fiber. Landraces grown from seed passed by farmers from one generation to the next, although adapted for survival in a particular region, may contain genes that enable them to resist environmental stresses, such as drought tolerance or disease resistance, in other regions.

Much of the corn in the Midwest is derived from two major races native to the Americas: the Northern Flints of the northern and eastern United States and southern Canada, and the Southern Dents derived from materials introduced from Mexico during the sixteenth century (Cox et al., 1988). By contrast, hard red winter wheat in the United States is largely derived from seeds brought by Mennonite immigrants in 1873 (Cox et al., 1988; Wilkes, 1988). Cultivated barley was introduced to the New World by early explorers, including Christopher Columbus (Cox et al., 1988). Soybean came from northern China, but an important introduction to the United States, designated PI 159925 (plant introduction number), arrived as part of an exchange of materials with Peru, where soybean is neither native nor a major crop.

Improving crop varieties from diverse backgrounds is a continuous process. Modern agriculture requires a flow of enhanced crop varieties that are productive in spite of pests, diseases, or climatic extremes. Improvements or changes in flavor, nutritive value, and shipping endurance have been added. Not long ago consumers sought out Golden Bantam sweet corn for their summer meals. Today, newer, more palatable varieties with improved disease resistance have replaced this long-time favorite. The highly productive agriculture of the Green Revolution was based on fertilizer-responsive dwarf varieties of wheat and rice assembled from widely divergent genetic materials (Dalrymple, 1986a,b). Each year presents the farmer, the consumer, and even the backyard vegetable gardener with numerous new varieties that may taste differently, are more productive, or have better disease resistance than previously released varieties. Perhaps not surprisingly, grass-roots, nonprofit groups have arisen to recover and protect the older cultivars of garden vegetables for a variety of conservation, cultural, and aesthetic reasons (Office of Technology Assessment, 1985; Shell, 1990).

Farm crops have always been vulnerable to environmental stresses, diseases, or pests, but farmers now grow crops that are more resistant to them. Diseases and pests change, however, and so must the varieties developed by plant breeders. New varieties must be developed to replace those endangered by shifts in pest and pathogen populations, or with outmoded performance or quality (see Figure 1-2) (National Research Council, 1972; Plucknett and Smith, 1982).

## Importance to Society and the Environment

Genetic resources can be lost or diminished through habitat destruction, displacement by other species, natural disasters, and neglect. As

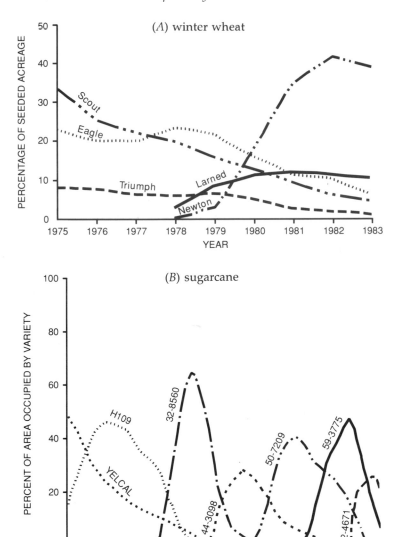

FIGURE 1-2 Commercial cultivars are often replaced with newly developed ones that have better production qualities, yields, or resistance to pests or pathogens, as illustrated in (A) winter wheat and (B) sugarcane. Sources: (A) Plucknett, D. L., N. J. H. Smith, J. T. Williams, and N. M. Anishetty. 1987. Gene Banks and the World's Food. Princeton, N.J.: Princeton University Press. (B) Plucknett, D. L., and N. J. H. Smith. 1986. Sustaining agricultural yields. BioScience 36(1):40–45. Reprinted with permission by (A) Princeton University Press, ©1987, and (B) American Institute of Biological Sciences, ©1986.

the basis for modern agriculture, they are strategic resources of concern to humanity. From these resources society derives food, shelter, clothing, pharmaceuticals, chemicals, and many other products.

The conservation of the broad range of plant genetic diversity, even when narrowed to agricultural concerns, is beyond the capacity of individuals, private companies, or small groups, although each of these can and does contribute to managing genetic resources. It requires the cooperation of a broad range of scientists, policymakers, administrators, and other concerned individuals. The secure and strategic conservation of genetic resources is a national government responsibility.

## International Exchange

The materials held in U.S. collections have provided important support to the agriculture of many nations. Some 110,000 accessions of the National Small Grains Collection are among the most frequently distributed genetic resources in the world. In response to requests from abroad, the United States dispatches more than 230,000 seed samples to over 100 countries each year.

The United States also plays a significant role in safeguarding important global collections. U.S. collections have been used to restore lost or damaged germplasm collections of other nations. For decades, the United States has served as a third-party quarantine site and supplier for developing nations seeking the germplasm of coffee, cocoa, rubber, and other species. Duplicate samples of rice from the International Rice Research Institute and maize from the Centro Internacional de Mejoramiento de Maíz y Trigo (International Maize and Wheat Improvement Center) are held as partial back-up collections for those institutions, to insure against loss. In addition, the United States cooperates with the International Board for Plant Genetic Resources (IBPGR) and the United Nations' Food and Agriculture Organization (FAO) to conserve the world's crop genetic resources. The United States has accepted responsibility for 18 of its collections, including those of maize, millet, rice, sorghum, wheat, beans (*Phaseolus* spp.), and soybean, to serve as international base collections within IBPGR's network (Hanson et al., 1984; International Board for Plant Genetic Resources, 1989; National Research Council, 1988).

## ORIGINS OF THE NATIONAL SYSTEM

The NPGS was established in 1974 as a collaboration of the federal government, states, and private industry to foster better management of the plant germplasm needed to sustain a productive agriculture.

Food in the United States is relatively inexpensive, abundant, varied, and safe, in part because of the success of modern plant breeding and agriculture. For example, new high-yielding, disease-resistant wheat varieties have ensured that flour has remained a relatively inexpensive, readily available commodity over many decades.

The NPGS has been described as a "user-driven system" (Murphy, 1988:210). It must serve the changing needs of a varied clientele in medicine, fiber, food, forage, industry, research, and other fields. It must also provide information and materials accurately and efficiently from its collections to those users.

## Development of Germplasm Activities in the United States

Efforts to gather genetic resources began in the early days of the Republic when various officials of the U.S. government asked citizens

The Institute of Plant Breeding in the Philippines maintains a sample of PI 314817. Credit: North Carolina State University, Peanut Breeding Laboratory.

**PEANUT**
PI 314817
*Arachis hypogaea* L.

GRIN Data

Cultivar: MANI
Origin: Peru
Acquisition: Peru
Maintenance
    site: Southern
    Regional Plant
    Introduction Station
Year PI assigned: 1966

It was serendipity that the peanuts collected by Dr. David Timothy in a trip down Peru's Huallaga River in 1966 turned out to confer resistance to two important diseases of the peanut. Timothy, a forage grass breeder at North Carolina State University, was in Peru to collect specimens of *Tripsacum*, a wild species related to maize. A storekeeper in the small river town of Juanjui, Province of Mariscal Caceres, in the Amazon basin gave him seeds of a locally grown peanut. The seeds were from remote plantings in sand bars on the Huayabamba River, up river from the village of Pachiza and above its confluence with the Huallaga River. Because the plants had been cultivated repeatedly in that area for a long time, Timothy guessed that they might possess distinct genetic traits. They were intro-

living or traveling abroad to send back seeds or plants of promising potential for new trees or crops (Hodge and Erlanson, 1956; Klose, 1950; White et al., 1989). John Quincy Adams, the sixth president, issued one such plea and similar requests came from Thomas Jefferson, Benjamin Franklin, and others.

From 1836 to 1862, before the U.S. Department of Agriculture (USDA) was established, the U.S. Patent Commissioner's office sent seeds and plants of foreign origin to farmers throughout the United States. The idea was supported by several members of Congress through the use of their postal franking privileges. This activity ended in 1923 when seed distribution became the responsibility of the USDA. It had, by then, grown from a cost of $1,000 in 1839 to $360,000 in 1922 (just under 11 percent of the USDA budget at that time).

In 1898, the Seed and Plant Introduction Section, which later became the Plant Introduction Office, was established to promote the exploration

---

duced into the United States and given the designation PI 314817.

It was subsequently discovered that these plants were resistant to two serious diseases, peanut rust and late leafspot. Peanut rust is especially devastating to crops in Central and South America, Africa, and Asia, where the crop is an important staple. Although not generally a problem in the United States, the disease can in some years cause crop losses of up to 70 percent in peanuts grown in southern Texas. Leafspot is more common in the United States and can reduce yields as much as 50 percent.

The genes for resistance to these diseases have been successfully transferred from PI 314817 into a breeding line of peanuts known as Tifrust-14, that was released cooperatively by the U.S. Department of Agriculture's Agricultural Research Service, the University of Georgia, and the International Crops Research Institute for the Semi-Arid Tropics in India. This line has been distributed by the U.S. Department of State's Agency for International Development in Thailand and in the Philippines to promote development of advanced breeding lines with rust resistance and high yields. This has been a lengthy process, and only recently have rust and leafspot resistant peanut lines been released to farmers. Several south African nations, Thailand, and the Philippines are also planning to grow the resistant varieties. The delays result from the time it takes to breed and select the few desired genes—in this case for disease resistance— while discarding unacceptable traits, such as poor quality or low yield. This deliberate, sometimes tedious, process of enhancement is crucial to harvesting the benefits hidden in an accession of plant germplasm.

---

*"GRIN Data" for the plant introduction (PI) number above represent information contained in the Germplasm Resources Information Network (GRIN). The narrative was prepared from information supplied by Johnny C. Wynne, Department of Crop Science, North Carolina State University.*

for and introduction of new crops. David Fairchild, Frank N. Meyer, and others introduced a broad range of new crops and genetic resources of existing crops (Cunningham, 1984; Hodge and Erlanson, 1956; Hyland, 1984; Klose, 1950; White et al., 1989). At the same time the plant introduction (PI) numbering system, still used by the NPGS, was established.

The section's Foreign Plant Introduction Office emphasized collecting. In the early 1930s, for example, H. G. MacMillan and C. O. Erlanson collected wild and primitive potatoes in Peru and Chile to obtain plants with genes for insect and disease resistance. Other plant hunters were in the West Indies searching for Sea Island cottons. R. Kent Beattie was completing a 5-year mission in China to collect chestnuts to replace the blight-stricken American chestnut. C. Westover and W. E. Whitehouse were searching for alfalfa in Russia as sources of resistance to bacterial wilt. Whitehouse then went on to Persia to collect melon, peach, apple, and pistachio germplasm. At the same time, W. J. Morse was making his now-famous contributions of thousands of wild and cultivated soybeans from China, Korea, and Japan.

The 1936 and 1937 editions of the *USDA Yearbook of Agriculture* recorded the genetic diversity of many crops of that time, but there were no nationally coordinated activities to preserve germplasm. In the 1940s the National Academy of Sciences' Committee on Plant and Animal Stocks expressed concern over the fate of the resources that formed the foundation for the world's crops. In a 1946 letter to Sir John Orr, the director general of the FAO, National Research Council Chairman Ross G. Harrison sought action by that organization and wrote the following (Harrison, 1946:1):

As improved varieties are introduced to production, large numbers of older, more diverse stocks disappear. A permanent loss of characters necessary for further improvement thus is likely to occur. As a safeguard to the welfare of all peoples, steps should be taken as soon as possible to collect and maintain the plant and animal materials likely to be of service in breeding.

In the summer of 1946 the 79th Congress passed Public Law 733; Title II of which is the Agricultural Marketing Act of 1946. This act provided the legal basis for establishing state-federal cooperation in managing crop and livestock genetic resources, and included an amendment to the earlier Bankhead-Jones Act of 1935 to support research (Title I). It also attempted to improve the marketing and distribution of agricultural products (Title II), and established a national advisory committee to the secretary of agriculture and the U.S. Department of Agriculture on matters of research and service authorized by the act (Title III).

The potato introduction project was begun in 1947 by breeders to maintain valuable South American and other potato germplasm; a site was established in Sturgeon Bay, Wisconsin, in 1950. The first of four regional plant introduction stations was established in Ames, Iowa, in 1948. Three more stations followed at Experiment, Georgia (now Griffin, Georgia); Pullman, Washington; and Geneva, New York. In 1958 the National Seed Storage Laboratory was opened at Ft. Collins, Colorado.

Following congressional passage of the Agricultural Marketing Act of 1946, action was taken to establish a cooperative enterprise that was coordinated through the newly designated Plant Introduction Section of the Agricultural Research Service (ARS). This included participation by ARS, the state agricultural experiment stations, the Cooperative State Research Service, and where appropriate, the Forest Service, the Soil Conservation Service, and the Bureau of Land Management. This combined effort involved close collaboration between those responsible for the acquisition and conservation of germplasm and federal, state, and private users. It was organized with a national office responsible for collecting and introducing germplasm, as well as through a series of regional and interregional stations responsible for the increase, maintenance, evaluation, documentation, and distribution of germplasm. Technical and administrative committees from the federal and state systems coordinated these activities.

**Emergence of the NPGS**

The NPGS has been described as a diffuse network of federal, state, and private institutions, agencies, and research stations (Council for Agriculture and Technology, 1984; General Accounting Office, 1981a,b; National Research Council, 1972; Office of Technology Assessment, 1987). Some reports have criticized its apparent inability to manage well all of the germplasm held by its cooperators. However, the system is still relatively young and is evolving to meet the rapid changes in technology and in the economic, legal, and political requirements of U.S. and world agriculture.

The present NPGS emerged 2 years after a 1972 restructuring of the ARS. The change underscored the recognition of the importance of genetic resources management and the need for a coordinated, national effort. The system has been an umbrella for an extensive array of germplasm management activities throughout the country.

U.S. scientists, in assembling a wide range of crop germplasm, presaged the recognition by other nations of the threat of loss of genetic diversity in primitive cultivars and landraces. Many NPGS stocks no longer exist where they were originally collected. Many early breeders' lines would also have been lost unless conserved in collections, such as those of the NPGS.

The NPGS is the world's largest distributor of plant germplasm. Accessions from its collections can be found in important crop collections throughout the world. Both public and private institutions have received plant materials, usually small amounts of seed, from the NPGS to support breeding and research programs. The NPGS does not supply seed for direct commercial use or agricultural production.

U.S. germplasm activities to a large degree have been based on an unofficial policy of ensuring national self-sufficiency. However, today the United States plays a more global role by providing support and leadership in planning and implementing international as well as U.S. programs.

Earlier commentaries on the NPGS have highlighted its shortcomings and needs as well as its successes (Council for Agricultural Science and Technology, 1984; General Accounting Office, 1981a,b; Office of Technology Assessment, 1987; U.S. Department of Agriculture, 1981). Increased appropriations, reapportionment of existing resources, construction of new facilities, and some centralization of responsibilities have been recommended to meet some of its more urgent needs (Office of Technology Assessment, 1987). Significant efforts have been made in recent years to address some of the concerns outlined in these reports and to improve the maintenance of germplasm held by the NPGS, but some issues persist.

This report does not review the findings of earlier reports, but analyzes and offers recommendations for future directions and activities of the NPGS. It proposes administrative, operational, and structural actions that could strengthen the NPGS and ensure a scientifically sound, responsive, and efficient system well into the twenty-first century. Toward this end, the components of the national system are described in the next chapter.

# 2

# Elements of the National Plant Germplasm System

The size and organization of a program to manage genetic resources varies with the goals and policies of a nation and the resources it is willing to commit to that purpose. There are, however, basic elements essential to all national programs for managing plant genetic resources. This chapter describes the components of the National Plant Germplasm System (NPGS) in the context of the elements of the model outlined in the preceding chapter (Figure 1-1).

## GENETIC RESOURCES IN THE UNITED STATES

The basic mission of the national system is to make available plant germplasm to scientists in the United States and worldwide for plant improvement, research, teaching, or extension programs. Activities include exploration, exchange, collection, and introduction; increase or regeneration; evaluation; documentation; preservation or maintenance; and distribution.

Prior to reorganization of the Agricultural Research Service (ARS) in 1972, the major components of the germplasm system were administered through the ARS New Crops Research Branch within the U.S. Department of Agriculture (USDA), or, for some major collections, through the specific USDA-ARS branch dealing with that commodity. In 1974, following the reorganization, germplasm activities were grouped into the National Plant Germplasm System to provide an umbrella system for germplasm acquisition, preservation, preliminary evaluation, and distribution. With the ARS holding lead administrative responsibility, the system was designed "to provide, on a continuing basis, the plant genetic diversity needed by farmers and public and private plant scientists to improve productivity of crops and minimize the vulnerability of those crops to biological and environmental stresses" (Jones and Gillette, 1982:1).

The NPGS is composed of several stations, repositories, or laboratories with varying responsibilities and locations throughout the United States (Table 2-1). The system holds more than 380,000 germplasm accessions representing more than 8,700 species.

While most of the activities of the U.S. germplasm program take place within the NPGS, no single site is solely responsible for all of them. The National Seed Storage Laboratory (NSSL), for example, is a specialized facility for long-term preservation that serves as a security backup to the active collections around the country. Seed viability testing and long-term storage are the major functions of the NSSL. Regeneration, characterization, evaluation, and distribution are performed elsewhere. While routine requests for seeds are often received at the NSSL, they are generally forwarded to and filled by a curator responsible for the particular active collection.

Cooperation and coordination among the components of the national system are essential. Individual sites may not be suitable for managing all of the accessions for which they have primary responsibility, and some accessions must be grown or regenerated at another site. Thus a station may maintain several species for which it has primary responsibility and may also grow plants for other collections. Materials that require short day-lengths to flower are problematic because the locations or other resources needed for regeneration are not always readily available.

Although spread over numerous locations, the NPGS is intended to function as an integrated national system for germplasm management. Plant materials entering or distributed from the NPGS follow predictable lines (Figure 2-1) of acquisition, conservation, management, and utilization. These activities are described below.

## ACQUISITION

The Plant Introduction Office (PIO) and its predecessors have been responsible for the acquisition of germplasm since 1898. Closely allied to this office are plant exploration activities and the health and quarantine of imported plant materials. Accession documentation begins when the PIO records passport data that accompanies new accessions for entry into the Germplasm Resources Information Network (GRIN) database. All of these activities are conducted through the National Germplasm

TABLE 2-1   Crop-Related Responsibilities in the National Plant
Germplasm System

| Facility and Location | Primary Crops or Species Conserved |
| --- | --- |
| Regional plant introduction station | |
| Western, Pullman, Washington | Common bean, garlic, *Allium* (onion) species, lupine, safflower, chickpea, wild rye, lettuce, lentil, alfalfa, forage grasses, horsebean, common vetch, milkvetch |
| Southern, Griffin, Georgia | Sweet potato, sorghum, peanut, pigeon pea, forage grasses, forage legumes, cowpea, mung bean, peppers, okra, melons, sesame, eggplant |
| Northeast, Geneva, New York | Tomato, birdsfoot trefoil, pea, clover, brassicas, onion |
| North-Central, Ames, Iowa | Maize, amaranth, oil-seed brassicas (e.g., rape, canola, mustard), sweet clover, cucumber, pumpkin, summer squash, acorn squash, zucchini, gourds, beet, carrot, sunflower, millets |
| National clonal germplasm repository | |
| Corvallis, Oregon | Filberts, pears, strawberry, raspberry, blackberry, cranberry, blueberry, mint, hops |
| Davis, California | Grape, stone fruits, walnut, almond, pistachio, persimmon, olive, fig, pomegranate, mulberry, kiwi |
| Geneva, New York | Grape, apple |
| Miami, Florida, and Mayaguez, Puerto Rico | Banana, mango, avocado, Brazil nut, Chinese date, jujube, coffee, cacao, soursop, bamboo, sugarcane, cassava, tropical yam, cocoyam |
| Orlando, Florida | Citrus |
| Hilo, Hawaii | Macadamia, guava, passion fruit, barbados cherry, breadfruit, jackfruit, pineapple, papaya, lychee, *Canarium* (pili nut), *Guiliema* (peach palm), *Nephelium* (rambutan, pulasan), carambola |
| Brownwood, Texas | Pecan, hickory, chestnut |
| Riverside/Brawley, California | Citrus and related genera, date |
| Other facilities | |
| National Arboretum, Washington, D.C. | Woody ornamental species |
| National Small Grains Collection, Aberdeen Idaho | Barley, oats, wheat, triticale, rye, rice, *Aegilops* (wild wheat relatives) |
| Interregional Research Project (IR-1), Sturgeon Bay, Wisconsin | Potato |
| Urbana, Illinois | Short-season soybean |
| Stoneville, Mississippi | Long-season soybean |
| College Station, Texas | Cotton |

```
┌─────────────────────────────────────────────────┐
│                 ACQUISITION                      │
│   Plant Introduction Office                      │
│   Germplasm Resources Information Network        │
│   National Plant Germplasm Quarantine Center     │
└─────────────────────────────────────────────────┘
                        │
                        ▼
┌─────────────────────────────────────────────────┐
│                CONSERVATION                      │
│   Regional plant introduction stations           │
│   National Small Grains Collection               │
│   National clonal germplasm repositories         │
│   Interregional Research Project, IR-1           │
│   Crop collections                               │
│   Genetic stock collections                      │
│   National Arboretum                             │
│   National Seed Storage Laboratory               │
│   Germplasm Resources Information Network         │
└─────────────────────────────────────────────────┘
                        │
                        ▼
┌─────────────────────────────────────────────────┐
│                 MANAGEMENT                       │
│   Regional plant introduction stations           │
│   National Small Grains Collection               │
│   National clonal germplasm repositories         │
│   Interregional Research Project, IR-1           │
│   Crop collections                               │
│   Genetic stock collections                      │
│   National Arboretum                             │
│   National Seed Storage Laboratory               │
│   Plant Introduction Office                      │
│   Germplasm Resources Information Network         │
└─────────────────────────────────────────────────┘
                        │
                        ▼
┌─────────────────────────────────────────────────┐
│                 UTILIZATION                      │
│   Scientists in colleges, universities, the federal│
│      government, and private industry            │
│   International exchange                          │
│   Germplasm Resources Information Network         │
└─────────────────────────────────────────────────┘
```

FIGURE 2-1  The division of responsibilities for germplasm management in the United States as they are coordinated through the National Plant Germplasm System.

Resources Laboratory (NGRL) at the Agricultural Research Service's Beltsville, Maryland, area.

Although plant exploration is an important part of the activities of the NPGS, the bulk of its collections have been acquired through exchange. Germplasm received from other collections worldwide accounts for an estimated 75 percent of acquisitions (S. Dietz, U.S. Department of Agriculture, personal communication, June 1989).

Not all of the germplasm that enters the United States comes in through the PIO. University researchers, botanical gardens and arboreta, companies, and private individuals all import plants and seed. Much of this material may not be duplicated or documented in NPGS collections.

Collections of unique germplasm, much of it privately held and not fully represented in the NPGS, form a considerable reservoir of diversity (Office of Technology Assessment, 1985). For example, groups such as the Seed Savers Exchange maintain heirloom or older varieties of vegetables, fruit, and flowers and distribute them primarily to individuals for personal use rather than for breeding new cultivars (Office of Technology Assessment, 1985; Shell, 1990). Their holdings are generally not part of NPGS collections and are not documented by it.

### Plant Introduction Office

The PIO is responsible for cataloging incoming germplasm accessions, assigning plant introduction (PI) numbers, and distributing new acquisitions to appropriate curators. The PIO publishes an annual inventory, listing the materials that have been assigned PI numbers (e.g., U.S. Department of Agriculture, 1988a,b), and coordinates germplasm exchanges.

The PIO may receive germplasm entering the United States for transfer to the appropriate NPGS site or record that a particular site has received such materials. All documentation for accessions is verified at the PIO. A PI number is assigned by the NGRL once passport data are verified. The PIO then distributes the material to appropriate sites.

The office also monitors some of the germplasm that enters the United States through avenues outside the NPGS (e.g., industry, botanical gardens, researchers), especially those plants or seeds receiving the attention of plant quarantine. If any of these latter materials are of importance to the NPGS, the importer is contacted by the PIO and invited to provide samples to the appropriate NPGS site or collection.

With the aid of the National Program Staff of the ARS, PI numbers are also assigned to new crop varieties, parental and advanced breeding

TABLE 2-2    Number of Plant Introductions into the United States, 1898–1987

| Date[a] | Number of Accessions Received | Number of Accessions in GRIN[b] | Percent | Percent of Total Received that Is Listed in GRIN[b] |
|---|---|---|---|---|
| 1898 to 1899 | 4,274 | 26 | 0.6 | 0.6 |
| 1900 to 1909 | 22,196 | 187 | 0.8 | 0.8 |
| 1910 to 1919 | 22,653 | 637 | 2.8 | 1.7 |
| 1920 to 1929 | 33,476 | 2,826 | 8.4 | 4.5 |
| 1930 to 1939 | 52,136 | 5,910 | 11.3 | 7.1 |
| 1940 to 1949 | 51,060 | 16,951 | 33.2 | 14.3 |
| 1950 to 1959 | 76,883 | 28,237 | 36.7 | 20.9 |
| 1960 to 1969 | 84,185 | 50,187 | 59.6 | 30.3 |
| 1970 to 1979 | 90,127 | 70,365 | 78.1 | 40.1 |
| 1980 to 1987 | 77,285 | 77,284 | 99.9 | 49.1 |

[a] A formal germplasm management effort began in the United States in 1948.
[b] Germplasm Resources Information Network.

SOURCE: Unpublished data supplied by the U.S. Department of Agriculture, Plant Introduction Office, July 26, 1987.

lines, and even genetic stocks that are registered by public and private plant breeders. Registrations for agronomic crops that document origin and important traits of a particular material may be published in *Crop Science* (Burgess, 1971; White et al., 1988).

Not all of the germplasm in the NPGS is assigned a PI number. Each collection site may maintain materials it received through channels other than that of the PIO. Currently, of the more than 372,000 accessions listed on the GRIN database about 70 percent are identified by PI numbers. Most of the balance carries identification numbers assigned by the NPGS sites.

Since 1898 more than 500,000 accessions have been received by the PIO (Table 2-2). Before the late 1940s, introductions went directly to interested researchers or breeders without any requirement that they be maintained beyond their usefulness to the individual. Duplicate or back-up samples were not held by the USDA since no facilities for that purpose existed until 1948. Consequently, most germplasm accessions obtained before 1948 are no longer available. However, many important breeder's lines and cultivars contain genes derived from them. Since 1948, an increasing proportion of germplasm introductions have become part of the NPGS collections and are listed on the GRIN (Table 2-2).

## Plant Exploration

Government-sponsored exploration for the purpose of collecting new germplasm is coordinated through the NGRL. The primary mission is the planning and implementing of plant explorations, especially in foreign areas (Purdue and Christenson, 1989). Plant exploration is a deliberate effort by the NPGS to seek and acquire specific kinds of germplasm.

Exploration proposals may be developed by one or more individual researchers who submit a formal proposal through the appropriate crop advisory committee, which is an NPGS advisory group specific to a crop. Scientists need not be employees of USDA or the national system to make such requests. Alternatively, proposals may be developed by a crop committee, the NGRL, or ARS National Program Staff. Once proposals are approved, qualified scientists undertake the exploration.

Some U.S. scientists conduct explorations using funds from other government sources, such as the National Science Foundation (NSF), or in cooperation with botanical gardens or arboreta. These activities

The grasses native to the Altai region of the south central Soviet Union are surveyed as part of an effort to collect wild rye species for U.S. germplasm collections. Credit: U.S. Department of Agriculture, Agricultural Research Service.

support botanical research, plant conservation, or other goals that may be unrelated to the perceived needs of the NPGS.

Exploration can also be accomplished through cooperation with individuals or organizations outside the United States. The NPGS occasionally cooperates with the International Board for Plant Genetic Resources (IBPGR) to collect materials of mutual interest. Cooperative exploration almost always involves scientists within the country where exploration occurs. In accordance with IBPGR practice, duplicate samples of collected germplasm are provided to germplasm collections within the country of origin.

### National Plant Germplasm Quarantine Center

When germplasm is acquired from other areas or regions, pests or pathogens may be introduced that could endanger domestic agriculture. Quarantine regulations are intended to reduce this risk, but they are not intended to facilitate the movement and entry of germplasm into the country. The role of expediting the quarantine process falls to the National Plant Germplasm Quarantine Center in Beltsville, Maryland, which recently was subsumed within the NGRL. The center represents a cooperative effort of the Animal and Plant Health Inspection Service (APHIS) and the ARS. ARS scientists and technical personnel provide expertise to propagate materials under quarantine and to test them for the presence of pathogens and other pests of quarantine significance. APHIS scientists and personnel certify such materials for release once quarantine regulations have been satisfied.

### Interregional Research Project-2

Obtaining pathogen-free, healthy plants for a number of clonally propagated crops, particularly fruit trees, can require years of quarantine. For this reason selected materials are maintained as virus-free clones that can be distributed without the delay associated with lengthy quarantine.

The Interregional Research Project-2 (IR-2), located in Prosser, Washington, functions as a national center for virus-free cultivars of deciduous fruit trees and selected ornamentals. It maintains more than 1,000 virus-free cultivars and holds the only known virus-free clones of a few germplasm materials. The IR-2 collections consist primarily of commercially important cultivars for use by researchers and industry. The program also develops methods for detecting and eliminating viruses from infected fruit tree cultivars. IR-2 maintains primarily the stocks of current importance for fruit production; it is not considered by many people to be a part of the NPGS.

## CONSERVATION

After acquisition by the NPGS, accessions are sent to the appropriate curator or collection for conservation. Conservation activities include increasing the sample size through grow-out if there are few seeds or plants, maintaining the material to preserve its genetic integrity, and ensuring that there is sufficient material for use.

In general, two kinds of conservation collections exist. Active collections, such as those at the regional plant introduction stations, multiply the material and are the primary sites for its distribution, characterization, evaluation, and general management. Base collections are back-up reserves of the active collections held under conditions of long-term storage. For seeds in the NPGS, base collections are held at the NSSL. For perennial woody plants, and some selected herbaceous species, which are usually propagated asexually, there are no back-up or base collections, and the national clonal germplasm repositories serve as the sites for both active and back-up collections.

### Active Collections

The central elements for managing and maintaining the germplasm of the NPGS are the many active collections throughout the United States. They are responsible for maintaining the germplasm, characterizing and evaluating it, and producing viable seed or planting materials, and they are the primary sources of material for distribution and exchange. The active collections of the NPGS include those of the regional stations and clonal germplasm repositories, and several commodity or special collections (Table 2-1). Other active collections, that are not part of the NPGS, exist in private or institutional collections at colleges, universities, and state agricultural experiment stations. Still other collections are held by industry, nonprofit organizations, botanical gardens, and arboreta. There is no precise information regarding the number, size, or condition of many of these mostly private collections, but it has been suggested that they probably represent a substantial germplasm pool (Office of Technology Assessment, 1985), some of which may be of considerable importance.

### Regional Plant Introduction Stations

Four regional stations have overall responsibility for maintaining the major seed-reproducing species held by the national system. These are the North-Central Regional Plant Introduction Station, Ames, Iowa; the Northeast Regional Plant Introduction Station, Geneva, New York; the Western Regional Plant Introduction Station, Pullman, Washington; and the Southern Regional Plant Introduction Station, Griffin, Georgia (Table 2-3). They are operated jointly by the ARS and state agricultural experiment stations through the Cooperative State Research Service (CSRS). Collectively, they hold approximately 135,000 accessions of nearly 4,000 species.

As originally envisioned in the 1940s and 1950s, the regional stations were established to meet the germplasm needs of plant breeders and other scientists. They were to provide foreign and native plant germplasm to crop scientists, preserve and evaluate introduced materials, and serve as holding facilities for the nation's genetic resources. Their responsibilities were based mainly on the concerns of agriculture in

A field area at Central Ferry, Washington, on the Snake River is operated by the Western Regional Plant Introduction Station and used for germplasm regeneration, evaluation, testing, and maintenance. Credit: U.S. Department of Agriculture, Western Regional Plant Introduction Station, Pullman, Washington.

each of their respective geographic regions. As each station was established, it was incorporated into the ongoing federal plant introduction
system, which in the late 1940s was headquartered at the Beltsville
(Maryland) Agricultural Research Center. Site selection was based in
part on accessibility, facilities, and by joint agreement of the directors
of the respective state agricultural experiment stations and the ARS. It
was agreed that the experiment stations would provide land, assist in
establishing laboratories, greenhouses, and related facilities, and provide
office space for staff. The ARS and CSRS supplied most of the funds
for equipment, operating expenses, and staff.

In addition to the regional stations, four federal plant introduction
stations were active during the 1940s and 1950s. Since then their status
has changed. The station at Glenn Dale, Maryland, served as an
introduction station and national quarantine center. It is being phased
out during the 1990s. The station at Miami, Florida, was responsible
for subtropical crops, particularly fruits, rubber, cacao, and coffee. One
of the earliest introduction stations, it has now become one of the clonal
germplasm repositories and continues to maintain its earlier collections.
The facility at Savannah, Georgia, held accessions of bamboo, sweet
potato, and several other crops, but it is no longer a federal plant
introduction station. Its germplasm was transferred to other sites. The
station at Chico, California, held germplasm of deciduous tree fruits
and nuts adapted to semi-arid conditions and has since been closed. A
portion of its accessions were transferred to clonal germplasm repositories.

The regional stations receive and distribute germplasm for most of
the species that can be stored as dry seed. Thus, they maintain the
active collections for much of the seed material in the national system.
They are responsible for seed increase and for depositing back-up
samples in the base collections of the NSSL. Curators at regional stations
interact regularly with users concerning management and use of the
species for which they have responsibility. Curators also characterize
and evaluate germplasm, but such activities can be limited by insufficient
funding or staff.

### National Clonal Germplasm Repositories

The clonal repositories (Table 2-4) contain active collections that hold
and propagate agriculturally important germplasm, such as strawberries,
raspberries, fruit trees, coffee, and nuts, that for a variety of reasons
are not usually held in active collections as seed. Eight repositories are
distributed over 10 sites, which together hold more than 27,000 acces-

TABLE 2-3 Overview of Holdings and Facilities at Sites and Collections, Excluding the National Clonal Germplasm Repositories

| Site or Collection | Year of Origin | Number of Species | Number of Accessions | Number of FTE Staff[a] | Facilities | | | |
|---|---|---|---|---|---|---|---|---|
| | | | | | Cold Storage (ft³) | Green-house (ft²) | Field Area (acres) | Lath or Screenhouse (ft²) |
| Regional plant introduction stations | | | | | | | | |
| Western | 1952 | 1,828 | 41,200 | 19.75 | 21,400 | 18,000 | 64 | 2,493 |
| Southern | 1949 | 1,320 | 58,031 | 23.0 | 16,000 | 16,409[b] | 8 | 3,108 |
| Northeast[c] | 1948 | 256 | 15,182 | 9.7 | 319 | 8,442 | 44 | 19,564[d] |
| North-Central | 1948 | 760 | 23,800 | 19.0[e] | 10,500 | 9,300 | 120 | 3,600 |
| National Small Grains Collection[f] | 1894 | 101 | 110,738 | 8.0 | 44,000 | 2,720 | 25 | |
| Interregional Research Project-1 (IR-1), potato | 1947 | 115 | 4,600 | 6.0 | 4,400 | 7,200 | 10 | 9,000 |

| | | | | | | | | |
|---|---|---|---|---|---|---|---|---|
| Soybean, long season | 1948 | 1 | 3,716 | 5.3 | 4,464 | 600 | 10 | 18,000 |
| Soybean, short season | 1949 | 14 | 9,440 | 4.5 | 450 | 300 | 8.5 | 1,500 |
| Cotton | 1955 | 40 | 5,558 | 3.0 | 2,723 | 3,672 | —[g] | 839 |
| National Seed Storage Laboratory | 1958 | 1,848 | 232,000 | 26.0 | 30,000 | 5,000 | 0 | 0 |

NOTE: Data are from 1988 and were supplied by individual sites, unless otherwise noted. N.A. = Not applicable.

[a] FTE staff = full-time equivalent staff.

[b] Includes about 12,000 square feet added during October 1989.

[c] A national clonal germplasm repository has separate facilities at this site, as described in Table 2-4.

[d] Includes barns, field laboratory, and seed cleaning, threshing, and drying spaces.

[e] Does not include 40 part-time, hourly positions.

[f] Data for 1988 were not available during the facility's relocation to Aberdeen, Idaho, from Beltsville, Maryland; the data were provided in 1989.

[g] Field areas are obtained through contract.

TABLE 2-4  Overview of Holdings and Facilities at the National Clonal Germplasm Repositories, 1988

| Repository Site[a] | Year of Origin[b] | Number of Species | Number of Accessions | Facilities | | | | |
|---|---|---|---|---|---|---|---|---|
| | | | | Number of FTE Staff[c] | Cold Storage (ft³) | Green-house (ft²) | Field Area (acres) | Lath or Screenhouse (ft²) |
| Brownwood, Texas | 1984 | 16 | 853 | 2.2 | 560 | 2,830 | 79 | — |
| Corvallis, Oregon | 1981 | 597 | 5,276 | 11.0 | 5,221 | 9,800 | 20 | 17,850 |
| Davis, California | 1981 | 142 | 6,572 | 6.0 | 3,500 | 2,400 | 70 | 5,200 |
| Geneva, New York[d] | 1984 | 60 | 3,700 | 4.8 | 2,300 | 1,800 | 50 | 900 |
| Hilo, Hawaii | 1987 | 35 | 535 | 4.25 | 118 | 1,800 | 10 | 3,000 |
| Orlando, Florida | 1988 | 31 | 400 | 1.0 | 0 | 1,400 | 10 | 1,800 |
| Miami, Florida | 1984 | 1,516 | 8,000 | 11.0 | 0 | 10,000 | 200 | 3,200 |
| Mayaguez, Puerto Rico | 1984 | 424 | 1,102 | 1.0 | — | 4,440 | 128 | — |
| Riverside, California | 1987 | 119 | 827 | 2.5 | 0 | 4,800 | 3 | 4,800 |
| Brawley, California | 1981 | 2 | 70 | 1.0 | 0 | 0 | 4.5 | 0 |

NOTE: The data were supplied by individual sites through a survey conducted in the fall of 1988.

[a] Mayaguez, Puerto Rico, is administered as part of the repository at Miami, Florida. Similarly, Brawley, California, is administered through Riverside, California.

[b] For Miami, Florida, and Mayaguez, Puerto Rico, the dates refer to the year in which they were designated as repositories. The sites originated in 1923 and 1901, respectively.

[c] FTE staff = full-time equivalent staff.

[d] The Northeast Regional Plant Introduction Station has separate facilities at this site, as noted on Table 2-3.

The characteristics of fruit on an apple germplasm accession are examined at the National Clonal Germplasm Repository at Geneva, New York. Credit: U.S. Department of Agriculture, Northeast Regional Plant Introduction Station, Geneva, New York.

sions. Accession decisions are generally made following consultation with public and private sector representatives, crop advisory committees or technical advisory committees, and ARS officials.

The primary responsibilities of the repositories are to collect, identify, propagate, preserve, evaluate, document, and distribute clonal germplasm as part of the NPGS. This includes maintenance of an information file on each accession in the clonal collection (National Plant Germplasm Committee, 1986). For most material held in such collections, long-term storage is not feasible, so duplicate materials are generally maintained in field and greenhouse or screenhouse collections to provide some back-up protection against loss. The repositories are charged with developing active global collections of appropriate wild species and domestic cultivars, and to assemble a maximum level of genetic diversity possible for each genus and species for which they are responsible. They also conduct research to improve evaluation, propagation, characterization, and preservation of clonal germplasm (National Plant Germplasm Committee, 1986).

The national clonal germplasm repositories are intended to carry out the same function for vegetatively propagated crops as that carried out by the regional stations and the NSSL for seed crops. Unlike seeds held at the laboratory, however, backups of clonal collections are limited to the same sites where the active collections are maintained. Many clonal crops can be conserved as seed, but they are impossible to maintain true to type by raising plants from seed. Many clonally propagated species take a long time to mature, and they are best preserved as mature live plants for plant breeding and research.

Clonal collections are expensive to establish, and they have many of the same problems that confront seed collections. Accessions must be maintained as plants in the field, which can require large tracts of land, or in screenhouses or greenhouses. Accessions may also be maintained as live sticks of budwood held under refrigeration or as tissue cultures. There may be losses during maintenance from insects and disease, freezing temperatures, electric power failures, or grazing animals. Clonal preservation is more expensive and labor intensive than seed storage. Clonal collections have been threatened as facility or land-use priorities have changed, as the principal scientists retired, died, or moved, or as funding declined. By establishing the national clonal germplasm repositories, a mechanism for stable, long-term maintenance for many important clonally propagated species has been provided.

## Interregional Research Project-1

The Interregional Research Project-1 (IR-1), located in Sturgeon Bay, Wisconsin, began in 1947 and is supported cooperatively by USDA through CSRS and ARS, and by the Wisconsin State Agriculture Experiment Station. It is the national repository for potato germplasm. The station has a collection of about 3,500 accessions of more than 100 wild and cultivated potato species and is an important global resource.

IR-1 uses a variety of methods to maintain germplasm. True seed of potatoes, in vitro plantlets of selected clones, and tubers are maintained. The germplasm is propagated both for maintenance and distribution.

## National Small Grains Collection

The National Small Grains Collection (NSGC) (Table 2-3) began in 1894 as a breeder's collection and is today the most widely used active collection in the NPGS (White et al., 1989). This collection, relocated in 1989 from Beltsville, Maryland, to Aberdeen, Idaho, holds more than 110,000 accessions of wheat, barley, oats, rice, rye, the wheat wild relative *Aegilops*, and the intergeneric wheat-rye hybrid, triticale. It is one of the world's largest collections with some of its accessions originating from nineteenth century plant explorations. About 100,000 samples are distributed yearly from the collection to breeders, researchers, and germplasm collections in the United States and abroad.

## Other Crop Collections

Important genetic resources are conserved and maintained at other federal, state, and university sites (Table 2-5). They include several crop

TABLE 2-5  Examples of Special Crop Germplasm Collections at State and Federal Facilities

| Crop | ARS (A) or State (S) Facility[a] | Location |
|---|---|---|
| Bamboo | A | Byron, Georgia |
| Barley[b] | S | Fort Collins, Colorado |
| *Brassica*[b] | S | Madison, Wisconsin |
| Cabbage[c] | S | Madison, Wisconsin |
| Cauliflower[c] | S | Madison, Wisconsin |
| Chickory[b] | A | Salinas, California |
| Clovers | S | Lexington, Kentucky |
| Cotton (*Gossypium barbadense*) | A | Phoenix, Arizona |
| Cotton (*G. hirsutum*) | A | Stoneville, Mississippi |
| Endive[b] | A | Salinas, California |
| Flax | A | Fargo, North Dakota |
| Forage and range grasses[b] | A | Logan, Utah |
| Forage grasses | A | Tifton, Georgia |
| Leafy vegetables | A | Salinas, California |
| Lettuce[b] | A | Salinas, California |
| *Linum* | A | Fargo, North Dakota |
| Maize[b] | S | Urbana, Illinois |
| Mustards[c] | S | Madison, Wisconsin |
| Native grasses | S | Brookings, South Dakota |
| Pea[b] | A | Geneva, New York |
| Pearl millet | A | Tifton, Georgia |
| *Pennisetum* (wild species) | A | Tifton, Georgia |
| Tobacco | A | Oxford, North Carolina |
| Tomato[b] and *Lycopersicon* species | S | Davis, California |
| Turnip[c] | S | Madison, Wisconsin |
| *Trifolium* species | S | Lexington, Kentucky |
| *Tripsacum* | S | Raleigh, North Carolina |
| Wheat and related species[b] | S | Columbia, Missouri; Riverside, California |
| Wild peanuts | A | Stillwater, Oklahoma |

[a] ARS = Agricultural Research Service.
[b] Primarily special genetic stocks.
[c] A breeder's collection that is largely duplicated at a regional station.

TABLE 2-6 Survey Results Showing Distinct Plant Genetic Resource Collections and Collections with Nine or More Accessions by Category at State Agricultural Experiment Stations

| State/ Territory | Number of Collections | Collections with Nine or More Accessions in a Category | | | | |
|---|---|---|---|---|---|---|
| | | Related Genetic Stocks | Wild Species | Land-races | Elite Breeding Lines | Cultivars |
| Alabama | 13 | 6 | 5 | 3 | 10 | 8 |
| Alaska | 3 | 2 | 0 | 0 | 0 | 3 |
| Arizona | 9 | 4 | 3 | 4 | 4 | 3 |
| Arkansas | 22 | 13 | 6 | 4 | 19 | 19 |
| California | 29 | 17 | 10 | 6 | 12 | 19 |
| Colorado | 15 | 6 | 2 | 7 | 5 | 5 |
| Connecticut | 5 | 2 | 2 | 0 | 2 | 0 |
| Delaware | 1 | 1 | 0 | 0 | 0 | 0 |
| Florida | 63 | 24 | 13 | 3 | 22 | 24 |
| Georgia | 18 | 11 | 5 | 4 | 14 | 13 |
| Hawaii | 24 | 17 | 10 | 5 | 10 | 19 |
| Idaho | 4 | 3 | 2 | 1 | 3 | 4 |
| Illinois | 14 | 10 | 4 | 4 | 10 | 10 |
| Indiana | 11 | 9 | 3 | 5 | 9 | 8 |
| Iowa | 20 | 8 | 4 | 3 | 12 | 11 |
| Kansas | 6 | 5 | 4 | 2 | 5 | 3 |
| Kentucky | 7 | 6 | 4 | 1 | 7 | 4 |
| Louisiana | 24 | 13 | 8 | 1 | 13 | 12 |
| Maine | 1 | 1 | 0 | 0 | 1 | 1 |
| Maryland | 5 | 0 | 1 | 1 | 3 | 3 |
| Massachusetts | 4 | 1 | 1 | 0 | 2 | 2 |
| Michigan | 18 | 10 | 10 | 2 | 14 | 7 |
| Minnesota | 28 | 16 | 8 | 5 | 24 | 18 |
| Mississippi | 4 | 1 | 0 | 1 | 2 | 2 |
| Missouri | 9 | 8 | 3 | 3 | 4 | 5 |
| Montana | 5 | 4 | 1 | 2 | 3 | 3 |
| Nebraska | 16 | 11 | 3 | 3 | 9 | 6 |
| Nevada | NR | NR | NR | NR | NR | NR |
| New Hampshire | NR | NR | NR | NR | NR | NR |
| New Jersey | 35 | 6 | 3 | NR | 17 | 21 |
| New Mexico | 7 | 7 | 3 | 3 | 5 | 4 |
| New York | 12 | 6 | 4 | 4 | 10 | 8 |
| North Carolina | 24 | 15 | 14 | 5 | 16 | 16 |
| North Dakota | 19 | 14 | 6 | 3 | 14 | 13 |
| Ohio | 10 | 9 | 2 | 2 | 7 | 8 |
| Oklahoma | 9 | 9 | 1 | 2 | 8 | 5 |
| Oregon | 21 | 9 | 5 | 2 | 12 | 12 |
| Pennsylvania | 13 | 9 | 2 | 1 | 10 | 4 |
| Puerto Rico | 6 | 4 | 0 | 4 | 6 | 3 |
| Rhode Island | NR | NR | NR | NR | NR | NR |

TABLE 2-6   (*Continued*)

| State/ Territory | Number of Collec- tions | Collections with Nine or More Accessions in a Category | | | | |
|---|---|---|---|---|---|---|
| | | Related Genetic Stocks | Wild Species | Land- races | Elite Breeding Lines | Cultivars |
| South Carolina | 7 | 6 | 2 | NR | 6 | 4 |
| South Dakota | 11 | 9 | 2 | 2 | 8 | 5 |
| Tennessee | 7 | 4 | 1 | 2 | 3 | 2 |
| Texas | 75 | 51 | 18 | 12 | 41 | 39 |
| Utah | 3 | 2 | 2 | NR | 2 | 2 |
| Vermont | NR | NR | NR | NR | NR | NR |
| Virginia | 5 | 3 | 3 | 0 | 4 | 4 |
| Washington | 12 | 7 | 5 | 0 | 11 | 7 |
| West Virginia | 1 | 0 | 1 | 0 | 0 | 1 |
| Wisconsin | 36 | 24 | 12 | 8 | 25 | 14 |
| Wyoming | 5 | 1 | 2 | 0 | 0 | 0 |
| Total | 696 | 403 | 200 | 121 | 424 | 384 |

NOTE: NR = No response.

SOURCE: Based on responses to a 1988 questionnaire distributed by the University of California Genetic Resources Conservation Program to plant science departments (including plant pathology, genetics, and agronomy, but not forestry) that were associated with state agricultural experiment stations.

species collections, such as soybeans and other legumes, cotton, and grasses. Some crop species, such as the native grasses in Brookings, South Dakota, or clovers in Lexington, Kentucky, are maintained in collections by individuals with particular interests in them or by sites with histories of studying them. Maintenance support is generally provided by the NPGS, or by state or other federal sources. Many of these collections are not duplicated and are primary sources for the materials they hold.

Several colleges, universities, and state agricultural experiment stations (Table 2-6) hold collections of germplasm that support breeding programs or the research interests of individuals. Much of this germplasm was gathered independently and may not be part of NPGS collections. Some of these collections are not listed in any official inventory, because they are small or only used by a few individuals.

The committee could not assess the degree to which these collections, often widely disseminated, duplicate NPGS collections. They may hold proportionately smaller numbers of wild species, landraces, and primitive materials than NPGS collections, but, as with the Charles M. Rick Tomato Genetics Resource Center (formerly the Tomato Genetics Stock Center), they can be important sources for wild or primitive materials.

### Genetic Stock Collections

This report defines genetic stocks as accessions with unique genetic or cytological characteristics that frequently make them of particular value in basic research. They do not include elite or advanced breeders' lines, in contrast with its definition by the Commission on Plant Genetic Resources of the Food and Agriculture Organization.

Genetic stocks differ from other germplasm stocks in several respects. Genetic stocks carry mutant genes or chromosomal rearrangements, deletions, or additions, and are often difficult and costly to maintain because they require specialized care and trained personnel. Stocks carrying recessive lethal traits must be maintained as hybrids. Cytological screening (chromosome examination) is needed to verify stocks with chromosomal abnormalities. Also, mutant stocks with lethal or deleterious genes often have short-lived seeds. These traits are often maintained in special genetic backgrounds or as groups of linked genetic markers.

Many genetic stock collections receive partial support from USDA through the ARS and from state sources. The NSF has been an important source of support, but it no longer funds maintenance of collections.

Genetic stock collections commonly held by the individuals or groups who developed them can be difficult to maintain because of their unique nature and frequently complex cytogenetics. Such collections are too easily lost when the person responsible for them moves on or retires. The NPGS has made efforts to determine where genetic stock collections are, to monitor them, and in some cases to provide funds through the ARS for maintenance.

The importance of genetic stock collections to crop development is well illustrated by the Charles M. Rick Tomato Genetics Resource Center at the University of California at Davis (Genetic Resources Conservation Program, 1988). The ARS provides partial support to the center, and an endowment fund has been established to support its work. It holds 2,750 accessions, about 1,000 of which are wild species related to the cultivated tomato. These wild species are important sources of resistance to 28 tomato diseases and pests (Table 2-7), as well as tolerance to other environmental stresses, such as temperature extremes, salinity, drought, and waterlogging. The tomato has also become a popular research model system in molecular studies of higher plant genomes. The detailed linkage map of the tomato genome was largely developed using the center's stocks.

TABLE 2-7 Economically Important Diseases of Cultivated Tomato to Which Wild *Lycopersicon* and *Solanum* Species Have Contributed Sources of Resistance

| Disease or Pest | Responsible Organism | Source of Resistance |
|---|---|---|
| **Fungi** | | |
| Collar rot | *Alternaria solani* | *L. hirsutum,* *L. peruvianum,* *L. pimpinellifolium* |
| Leaf mold | *Cladosporium fulvum* | *L. esculentum* var. *cerasiforme* |
| Anthracnose | *Collectotrichum coccodes* | *L. esculentum* var. *cerasiforme* |
| Target leaf spot | *Corynespora cassiicola* | *L. pimpinellifolium* |
| Didymella canker | *Didymella lycopersici* | *L. hirsutum* |
| Fusarium wilt | *Fusarium oxysporum* f. sp. *lycopersici* | *L. pimpinellifolium* |
| Phoma blight | *Phoma andina* | *L. hirsutum* |
| Late blight | *Phytophthora infestans* | *L. pimpinellifolium* |
| Phytophthora fruit rot | *Phytophthora parasitica* | *L. pimpinellifolium* |
| Phytophthora root rot | *Phytophthora parasitica* | *L. esculentum* var. *cerasiforme* |
| Corky root | *Pyrenochaeta lycopersici* | *L. peruvianum* |
| Septoria leaf spot | *Septoria lycopersici* | *L. esculentum* var. *cerasiforme,* *L. hirsutum,* *L. pimpinellifolium* |
| Gray leaf spot | *Stemphylium solani* | *L. pimpinellifolium* |
| Verticillium wilt | *Verticillium albo-atrum* | *L. esculentum* var. *cerasiforme* |
| Dahlia wilt | *Verticillium dahliae* | *L. peruvianum* |
| **Bacteria** | | |
| Bacterial canker | *Corynebacterium michiganese* | *L. hirsutum,* *L. peruvianum,* *L. pimpinellifolium* |
| Bacterial speck | *Pseudomonas tomato* | *L. pimpinellifolium* |
| Bacterial wilt | *Pseudomonas solanacearum* | *L. pimpinellifolium* |
| Bacterial spot | *Xanthomonas vesicatoria* | *L. esculentum* var. *cerasiforme* |
| **Nematodes** | | |
| Potato cyst nematode | *Globodera pallida* | *L. hirsutum* |
| Sugarbeet nematode | *Heterodera schactii* | *L. pimpinellifolium* |
| Root-knot nematode | *Meloidogyne incognita* | *L. peruvianum* |

*(continued)*

TABLE 2-7 (*Continued*)

| Disease or Pest | Responsible Organism | Source of Resistance |
|---|---|---|
| Viruses | | |
| Cucumber mosaic | Cucumber mosaic virus | *L. peruvianum,*<br>*S. lycopersicoides* |
| Curly top | Beet curly top virus | *L. peruvianum* |
| Veinbanding mosaic | Potato Y virus | *L. esculentum* var.<br>*cerasiforme* |
| Spotted wilt | Tomato spotted wilt<br>virus | *L. pimpinellifolium* |
| Tobacco mosaic | Tobacco mosaic virus | *L. peruvianum* |
| Tomato yellow leaf | Tomato yellow leaf<br>curl virus | *L. cheesmanii,*<br>*L. hirsutum,*<br>*L. peruvianum,*<br>*L. pimpinellifolium* |

SOURCE: Rick, C. M., J. W. DeVerna, R. T. Chetelat, and M. A. Stevens. 1987. Potential contributions of wide crosses to improvement of processing tomatoes. Acta Horticulturae 200:45–55.

Some genetic stock centers are expanding to include the maintenance of cloned DNA (deoxyribonucleic acid) sequences for use as molecular markers (restriction fragment length polymorphisms) and as specifically cloned genes. As these technologies develop and are increasingly used in research and breeding, the NPGS will have to consider how best to manage the data and DNA.

## National Arboretum

The National Arboretum, established in 1927 in Washington, D.C., by an act of Congress, is primarily concerned with research and public education on trees, shrubs, and related plants. It maintains its own herbarium (separate from the National Herbarium of the Smithsonian Institution) of more than 500,000 dried specimens with emphasis on economic and cultivated plants, including voucher specimens from USDA plant explorations. It cooperates closely with NPGS on an informal basis. An ARS plant germplasm staff person at the arboretum has the primary responsibility of developing a germplasm collection of woody landscape species.

The arboretum has an active worldwide germplasm collection program and distributes materials to users in both the public and private sectors.

It maintains about 60,000 accessions, primarily of ornamental trees and shrubs. Most are conserved as plants in outdoor plantings and in screenhouses or greenhouses. The arboretum has introduced many widely used ornamental plants. For example, the popular ornamental crapemyrtle hybrids, resistant to the powdery mildew fungus, were developed and introduced by the National Arboretum.

## Base Collections

Some NPGS genetic resources are also held in base collections in long-term storage as a reserve to the active collections. If samples in active collections are no longer available or cannot be regenerated, the germplasm held in the base collection can be used to replace them.

Dried specimens of woody plants are among the materials held in the herbarium of the National Arboretum, Washington, D.C. Credit: U.S. Department of Agriculture, Agricultural Research Service.

Germplasm in a base collection must be maintained under conditions that promote long-term storage. For seeds this generally entails maintenance at low temperatures and low relative humidity. It may also include cryopreservation (storage in or suspended above liquid nitrogen between −150°C and −196°C) of seeds, pollen, in vitro cultures, or dormant buds, in the case of clonal germplasm.

## National Seed Storage Laboratory

The NSSL is the principal site of long-term seed storage of genetic resources in the United States (see Table 2-3). It provides base collection storage facilities for the national system and holds samples of other important seed accessions (Table 2-8). Consideration is being given to storing cell and tissue cultures, pollen, and DNA.

*Specimens of* Lagerstroemia fauriei *can be found at the National Arboretum, Washington, D.C. Credit: U.S. National Arboretum.*

### CRAPEMYRTLE
PI 237884
*Lagerstroemia fauriei*
Koehne

GRIN Data

Origin: Unknown
Acquisition: Japan
Common name:
  Crapemyrtle
Year PI assigned: 1957

A wild species (*Lagerstroemia fauriei*), related to the popular ornamental tree, crapemyrtle (*L. indica*), illustrates the importance of wild plants as germplasm for improving and protecting horticultural plants.

On November 26, 1956, when Dr. John Creech collected seeds of *L. fauriei* from its native habitat near a rocky streambed at an elevation of 1,000-feet (300 meters) on the small Japanese island of Yakushima, the common crapemyrtle was already

The NSSL currently holds in excess of 230,000 accessions of more than 1,800 species. It also holds reference specimens of germplasm registered through the Crop Science Society of America, and it is the sole site for some genetic stock collections such as the Jimson weed (*Datura*) collection of A. F. Blakeslee, important to early studies of chromosomes and plant development. Seed samples of plant varieties protected under the Plant Variety Protection Act are also held by the NSSL.

Selected collections of the national system have been designated by the IBPGR as part of its international network of base collections (Table 2-9). These may be exchanged as a part of the international collaboration between the NPGS and IBPGR. Many of these collections existed before the IBPGR designation.

The NSSL provides back-up storage for some major international

---

a popular ornamental in the United States. Creech was attracted to this particular relative of the horticultural plant by what he described as cinnamon-colored bark, which peeled away from the trunk. But its most important characteristic was its later-discovered genetic resistance to powdery mildew (*Erysiphe lagerstroemiae*), a disease which disfigures the leaves and flowers of infected cultivated plants, particularly in the humid south.

Seed of *L. fauriei* brought back by Creech was assigned a plant introduction number and initially grown at the Plant Introduction Station, Glenn Dale, Maryland. From there seedlings were widely distributed to botanical gardens and arboreta in the United States, including the National Arboretum in Washington, D.C., where it was tested for susceptibility to powdery mildew and found to be resistant.

Dr. Donald Egolf, a research horticulturist at the National Arboretum, crossed *L. fauriei* with the common crapemyrtle to produce mildew-resistant hybrids. These new hybrids, some of which display novel flower colors, are rapidly replacing the older common crapemyrtle.

Meanwhile, the cinnamon-colored bark that originally attracted Dr. Creech to *L. fauriei* has made it a desirable ornamental in its own right. Trees can now be found at many botanical gardens and arboreta in the South. Ironically, in its native habitat the tree is threatened by intensive forestry and cutting for charcoal.

---

*"GRIN Data" for the plant introduction (PI) number above represent information contained in the Germplasm Resources Information Network (GRIN). The narrative was prepared from information supplied by John L. Creech, U.S. Department of Agriculture (retired).*

TABLE 2-8  Back-up Storage at the National Seed Storage Laboratory (NSSL)

| Site/Collection | Total Accessions | Number Held by NSSL | Percentage of Total Accessions |
|---|---|---|---|
| Site | | | |
|   Ames, Iowa | 29,732 | 8,833 | 30 |
|   Geneva, New York | 15,451 | 8,243 | 53 |
|   Griffin, Georgia | 59,290 | 15,993 | 27 |
|   Pullman, Washington | 40,834 | 23,625 | 58 |
| Collection | | | |
|   Clover | 271 | 189 | 70 |
|   Cotton | 4,949 | 2,248 | 45 |
|   Flax | 2,659 | 2,378 | 89 |
|   National Small Grains Collection[a] | 114,480 | 77,927 | 68 |
|   Soybean | 12,807 | 10,776 | 84 |
|   Tobacco | 1,272 | 667 | 52 |

[a] A small sample of every accession in the National Small Grains Collection (NSGC) was sent to the NSSL when the collection was moved from Beltsville, Maryland, to Aberdeen, Idaho, in 1989. These were not of sufficient size to constitute back-up samples and additional seed is being produced. All barley and oats from the NSGC are also backed up at the Canadian National Gene Bank, Ottawa, Ontario.

SOURCE: Data supplied by the Database Management Unit, Germplasm Services Laboratory, November 30, 1989.

collections. For example, it holds accessions of rice from the International Rice Research Institute (National Research Council, 1988), but these are not inventoried as part of the NPGS collection. The seed accessions are maintained in bond, under quarantine in their original shipping cartons. Their disposition is at the discretion of the originating institution. The NPGS deposits duplicates of its barley and oat collections with the National Plant Gene Conservation Center in Ottawa, Canada, and is seeking arrangements with other nations to hold backups of other collections (H. L. Shands, U.S. Department of Agriculture, personal communication, September 1989).

As part of its comprehensive seed storage function, the NSSL periodically tests seeds to ensure continued viability. Regional stations, curators, and other cooperators are expected to regenerate supplies of seed as requested by the NSSL, although additional funds are not provided for this purpose. The NSSL does not regenerate, evaluate,

Back-up accessions of the collection of the International Rice Research Institute are kept in subfreezing storage at the National Seed Storage Laboratory.

enhance, or distribute germplasm as part of its mandate. The laboratory has begun efforts to rescue valuable seed samples of low viability. At present there are no effective plans for regenerating certain seed collections held at NSSL that are unadapted to conditions anywhere in the continental United States or Puerto Rico, even though the NPGS has accepted international base collection responsibility for such collections. This underscores the need for additional facilities and the impor tance of cooperating with other countries in the regeneration of such germplasm.

The present buildings of the NSSL were completed in the late 1950s and are still structurally sound. However, technological advances in seed storage and the need for more storage space have made the construction of a new, enlarged, and modernized laboratory imperative. Recommendations for the design of an expanded NSSL facility are given in a separate committee report (National Research Council, 1988). Until such facilities are completed, space will continue to be a major constraint on NSSL activities.

### Vegetatively Propagated Germplasm

Germplasm held by the national clonal germplasm repositories and the IR-1 is not often stored in base collections as seed. In general, there

TABLE 2-9   Collections of the National Plant
Germplasm System Designated by the International
Board for Plant Genetic Resources (IBPGR) as
International Base Collections

| Collection | IBPGR-Designated Scope |
|---|---|
| Maize | New world |
| Millets (*Pennisetum*) | Global |
| Rice | Regional |
| Sorghum | Global |
| Wheat | Global |
| Beans (*Phaseolus*, cultivated species) | Global |
| Soybean | Global |
| Vigna (*Vigna unguiculata*) | Global |
| Sweet potato | Global |
| Allium | Global |
| Amaranthus | Global |
| Okra | Global |
| Tomato | Global |
| Cucurbits (*Cucumis, Citrullus, Cucurbita*) | Global |
| Eggplant | Global |
| Sugarcane | Global |
| Forage legumes (*Leucaena, Zornia*) | Global |
| Forage grasses (*Cynodon, Pennisetum, Paspalum*) | Global |

SOURCE: International Board for Plant Genetic Resources. 1989.
Annual Report 1988. Rome, Italy: International Board for Plant
Genetic Resources.

are no back-up or base collections of clonally propagated species,
although clonal repositories typically maintain duplicates of field-grown
accessions in screenhouses or greenhouses. For some species, seed
storage is possible and some of these are held as part of the NSSL base
collections. However, seeds from a heterozygous, clonally propagated
plant, while preserving the genes of the accession, do not preserve the
specific gene combinations that determine the plant's precise characteristics. It may be desirable to conserve selected gene combinations for
some species, such as those that require several years to reach maturity.

For materials that must be maintained as clones, the active collections

Pineapple germplasm at the National Clonal Germplasm Repository, Hilo, Hawaii, is maintained *(A)* in the field and *(B)* as in vitro laboratory cultures. Credit: U.S. Department of Agriculture, National Clonal Germplasm Repository, Hilo, Hawaii.

usually keep back-up samples as field plantings or in screenhouses or greenhouses. Back-up accessions also may be held as in vitro tissues, plantlets, or other propagative material (e.g., budwood) held in a quiescent state, under special conditions. These back-up holdings are not regarded as base collections because their storage is neither long term nor secure from environmental loss (particularly for field-maintained duplicates). However, they provide some security against complete loss of the primary active collection. As tissue culture and cryopreservation become more reliable as preservation methods, it may be possible to provide back-up holdings of clonal materials at sites separate from the active collection. For example, the NSSL maintains a set of cryopreserved strawberry cultures as an experimental backup to the field collection at the repository in Corvallis, Oregon.

## MANAGEMENT

Germplasm management must ensure the continued high viability of all accessions and prompt, accurate distribution of samples in response to requests. Information about accessions, including passport, characterization, and evaluation data, must be in an accessible form to enable users to select from a large array of materials. This is accomplished in the national system through the cooperation of several sites, and it is facilitated by the GRIN database. The elements of this cooperation include the testing of seed viability, increase (regeneration) of seed and clonal germplasm, and distribution.

### Seed Viability Testing

Germplasm maintained as seed must be tested regularly to monitor viability. This is carried out by the responsible NPGS site or in cooperation with the NSSL.

Seed germination tests under controlled conditions are performed at intervals that vary with the physiology of the species. Lettuce (*Lactuca*) and onion (*Allium*), for example, are more short-lived than many cereal grains or beans and need more frequent testing. Chemical tests of seed viability are less reliable and not generally practiced at NSSL; there is no reliable technique for the nondestructive assessment of viability.

Accessions received with low seed viability pose several difficulties, in part because seed with low viability may not survive well under storage (Priestley, 1986). If the sample size is small, prompt and careful regeneration must be performed to avoid further loss of genetic diversity.

The NSSL and IBPGR standards for viability at one time were different.

Seed is packaged and sealed in foil-laminated moisture-proof bags for storage at the National Seed Storage Laboratory. Credit: U.S. Department of Agriculture, Agricultural Research Service.

IBPGR set a standard of 85 percent germinability for seed, while the NSSL's minimum standard was 60 percent, but it was recently raised to match IBPGR. For a few accessions, such as some chromosomally aberrant genetic stocks, a lower standard may be appropriate because of their inherently low seed viability, even in freshly grown seed. Current viability data from the NSSL are based on a germination rate of 65 percent. Table 2-10 shows that 164,603 accessions (71 percent) in the collection have a germination rate greater than 65 percent. However, 45 percent of the accessions contain less than 550 seeds.

TABLE 2-10 Germination Rate and Number of Seeds for Accessions at the National Seed Storage Laboratory (NSSL)

| Total Number of Seed in Accession | Germination Rate Per Accession | | | Total Accessions | Percentage of All Accessions |
|---|---|---|---|---|---|
| | >65 Percent | <65 Percent | Unknown | | |
| More than 550 seeds | 114,534 | 13,558 | — | 128,092 | 55 |
| Less than 550 seeds | 50,069 | 5,042 | 49,007 | 104,118 | 45 |
| Total | 164,603 | 18,600 | 49,007 | 232,210 | |
| Percentage | 71 | 8 | 21 | | |

SOURCE: S. Eberhart, NSSL, personal communication, February 8, 1990. Data are based on germination tests conducted between 1979 and 1989.

More important than viability, however, is whether the aged seed survives as a consequence of the genes it possesses. Over time selection may favor the survival of particular genotypes. Clearly, viability testing, sampling, and rejuvenation policies and procedures are critical to maintaining the integrity of accessions.

## Regeneration and Multiplication

Viability tests, seed distribution, and the decline of viability in storage all reduce the numbers of viable seeds of an accession. New accessions with few or poorly viable seeds require immediate increase. Increase may be necessary when a large amount of seed is needed for special evaluations. Seed regeneration or increase is normally the responsibility of those holding active collections, such as the regional stations, but it may be requested either from the scientist who contributed the materials

Germplasm accessions of chickpea, lentils, and broad bean are grown to produce fresh seed at the Western Regional Plant Introduction Station. Credit: U.S. Department of Agriculture, Western Regional Plant Introduction Station, Pullman, Washington.

or from a third party, often on an unsupervised contract or cooperative basis.

Regeneration techniques should be designed to minimize changes in the relative abundance (frequencies) of genes within an accession. Under the extremes of drought, cold, or disease, for example, some individuals in an accession could survive or even thrive, while others would die, resulting in changes in the overall genetic composition of the sample (i.e., genetic shift). Because of the small population size of an accession, random genetic change (i.e., genetic drift) can also have a major effect. To prevent these genetic changes, parental populations of 100 to 200 plants are grown in an environment to which they are adapted to avoid unintended selection. Other protocols call for the use of multiple sites in varying environments or over succeeding years. Nonetheless, genetic changes may occur during the replenishment of aging seed through inadvertent selection during regeneration and seed aging (Roos, 1984a,b). Methods of storage and testing that reduce the frequency of regeneration are particularly important.

No procedure ensures that samples returned from regeneration have the same genetic composition as those that were sent out. New techniques, such as the comparison of gene frequencies before and after regeneration using restriction fragment length polymorphisms, or older technologies such as electrophoretic analysis of seed proteins, could provide some measure of assurance. The common practice is to assume that seed for regeneration has been grown correctly and protected from genetic change or contamination.

Clonal crops have specific requirements for germplasm maintenance and increase. Temperate tree fruits may require isolation to prevent loss due to pests or diseases. Screened area protection, quarantine isolation, and standard orchard management practices are among the methods employed. Specific environments may also be needed to ensure the normal development of plants or fruit. Herbaceous plants, such as potatoes, may require greenhouse culture, specific propagation methods, or unique cultural practices. As a consequence, clonal repositories can be very costly.

## Distribution

Distribution of germplasm to researchers and breeders in the United States and throughout the world is an important part of the activities of the national system (Table 2-11). Germplasm requests to the NPGS from outside the United States (except those from Canada) are processed through the Plant Introduction Office. Distribution for U.S. and Cana-

TABLE 2-11 Germplasm Distributions from Seed Collections and Clonal Repositories of the National Plant Germplasm System to Various Users, 1980–1989

| Year | United States | | | USAID[a] | Foreign Private[a] | Foreign Public[b] | International Centers | Annual Total |
| | Private | State | Federal | | | | | |
|---|---|---|---|---|---|---|---|---|
| 1980 | 14,349 | 57,463 | 60,197 | 660 | 3,449 | 35,649 | 4,880 | 176,647 |
| 1981 | 13,856 | 49,491 | 47,104 | 85 | 3,788 | 65,269 | 8,774 | 188,368 |
| 1982 | 21,086 | 55,431 | 43,982 | 116 | 2,332 | 47,350 | 7,481 | 187,778 |
| 1983 | 12,355 | 63,907 | 76,768 | 2,238 | 3,948 | 36,269 | 4,715 | 200,200 |
| 1984 | 17,596 | 82,787 | 175,838 | 249 | 3,855 | 20,313 | 2,915 | 309,553 |
| 1985 | 20,941 | 103,558 | 94,743 | 578 | 1,985 | 25,874 | 1,345 | 249,024 |
| 1986 | 13,787 | 49,502 | 76,331 | 388 | 3,658 | 16,697 | 3,546 | 163,909 |
| 1987 | 11,471 | 39,668 | 86,302 | 216 | 13,154 | 28,938 | 3,409 | 183,158 |
| 1988[c] | 18,736 | 44,934 | 56,656 | 56 | 3,579 | 22,176 | 3,348 | 149,485 |
| 1989 | 28,575 | 86,306 | 84,909 | 2,590 | 5,253 | 14,882 | 7,809 | 230,324 |

[a] Distributions were made through the U.S. Agency for International Development (USAID) to 36 countries in 1987.

[b] Distributions were made to 104 countries in 1987.

[c] Distributions for 1988 declined because distributions from the National Small Grains Collection were deferred while the collection was being moved from Beltsville, Maryland, to Aberdeen, Idaho.

SOURCE: Unpublished data supplied by U.S. Department of Agriculture. Plant Introduction Office, April 10, 1990.

dian requests is done by the appropriate collection. A considerable amount of germplasm, particularly genetic stocks and breeding lines, is directly exchanged between scientists.

The PIO coordinates most of the international transfer of germplasm by the national system. In cooperation with the responsible active collection or curator, the PIO staff might work to determine which materials are appropriate to fill a request too large to be of practical use (e.g., a request for all of the accessions of a crop that may number in the thousands). Where quarantine regulations require certification or declaration of phytosanitary status of the material, the PIO, in cooperation with APHIS, completes the needed paperwork. This is frequently accomplished at the local level by qualified state or federal personnel.

Germplasm is not distributed from the NSSL unless it is unavailable from active collections or elsewhere in the NPGS. In this case, the NSSL generally sends the seed to an appropriate active collection site or the original supplier of the germplasm for regeneration. In cases of extreme need or emergency and if sufficient seed is available, the NSSL may distribute samples directly to users. If it is necessary to regenerate samples, distribution of requested germplasm can be delayed for months or, more rarely, years.

The number of seeds distributed depends on the species and the amount available in storage. In general enough seed is sent for the user to grow a row approximately 15 meters (50 feet) long. From this, requesters are expected to reproduce additional seed. Individuals requesting seed for personal use, as in gardens, are generally referred to other sources. Recipients of NPGS germplasm are asked to acknowledge its source in reports and publications, and to report performance or evaluation information to the NPGS curator for that crop.

### Characterization and Evaluation

To be useful and accessible to breeders and researchers, information about the germplasm in a collection should be well described for specific characters and should indicate origin. It will enable users to know which accessions are likely sources for particular genetic traits. The maintenance of this information in a retrievable form is referred to as documentation.

Passport data provide the basic documentation, including taxonomic designation and information about where an accession was collected. These data are provided by the collector or donor institution. They are reviewed upon arrival at the PIO and entered into the database record when a PI number is assigned by that office. Unfortunately, such minimal data are not readily available for many of the accessions of the

NPGS, particularly those received before 1978. While passport data do exist for many accessions, the NPGS has been slow in adding this information to its computerized database.

Characterization involves the assessment of a varying number of descriptors ranging from morphological to biochemical. These descriptors, intended to describe an accession in relation to others in the collection, are determined by crop advisory committees and curators. The information can be gathered by the curators of active collections as materials are regenerated or assessed by cooperating scientists during evaluation for resistance to disease, environmental responses, or other traits. The gathering of crop descriptor data is an important part of the work of regional stations, repositories, and commodity collections because such data better define germplasm holdings and aid re-identification during regeneration.

Evaluation is a lengthy, often repetitive process of examining accessions for traits of significance to potential users. Screening for disease resistance is an example. It may take many years to test all of the accessions for a very large collection in a sufficiently wide range of environments. Although preliminary evaluation for genetically stable traits is generally considered an NPGS activity, more detailed evaluation for characters such as disease resistance or production qualities is generally part of the research that accompanies a breeding program. In a few cases, such as alfalfa at the western regional station, NPGS sites have funded researchers at other locations to conduct evaluations. Private companies and other users of NPGS germplasm also evaluate material, and the NPGS has occasionally provided funds for evaluating certain traits.

The Soil Conservation Service (SCS) also cooperates with the NPGS through a memorandum of understanding with the ARS to perform evaluations on certain materials. One purpose of the SCS is to foster acquisition, evaluation, and distribution of plants important to soil and water conservation.

## Documentation

The usefulness of new accessions depends on the user being able to retrieve information about them. In 1977 ARS recommended setting up a central repository for genetic resources information with standardized crop descriptors that would improve the management of NPGS collections (Mowder and Stoner, 1989). The Germplasm Resources Information Network manages all of the data associated with acquisition, evaluation, regeneration, inventory, and other records of the NPGS collections. It is a centralized computer database for managing and operating the national system and for informing scientists and other users about the location and characteristics of NPGS germplasm (Database Management Unit, 1987).

The GRIN runs on a Prime 9955 Model II computer and uses a software package and database management system distributed by Prime Computer, Inc. Additional programs are written to meet specific needs. This system is transportable only to other, compatible Prime computers. Smaller portions of the data can be formatted for use on microcomputers using commonly available software packages. Searches of the database with printed results can be requested.

The GRIN holds three kinds of information. Passport data are recorded by the PIO when the PI number is assigned. They include the name of the collector, collection site data, taxonomy, and collection longitude, latitude, and elevation. Characterization and evaluation data are of value to potential users and include general plant descriptions, agronomic responses, disease and insect pathogen susceptibility or resistance, quality, and yield. Finally, the GRIN provides inventory and seed request processing data to NPGS sites as an aid to managing their collections.

While administering the network is the responsibility of the GRIN staff in Beltsville, Maryland, most sites also have personnel trained to enter their evaluation and other site data directly into the network.

## UTILIZATION

The NPGS centers, for the most part, do not develop new cultivars or improve breeding materials. Nonetheless, a considerable amount of public funds are dedicated to germplasm enhancement, or the transfer of useful traits into an agronomically appropriate genetic background. For some commodity collections at experiment stations or universities, germplasm conservation may be an adjunct to enhancement.

Many public institutions are involved in germplasm enhancement. State agricultural experiment stations, public and private colleges and universities, and private industry use NPGS collections for breeding and enhancement to produce improved parental materials. Breeders do not use these collections frequently, but find them invaluable for particular genetic characters not readily available to them (Duvick, 1984; Peeters and Galwey, 1988). For example, the appearance of the Russian wheat aphid in the United States in 1986 caused more than $130 million in losses to grain crops in 1988 (Peairs et al., 1989). An immediate search for host plant resistance in the national wheat collection has already revealed some promising sources of resistance.

Russian wheat aphids feed on wheat leaves, which then roll and impede the effectiveness of conventional pesticide applications. Genetic resistance to the aphid is found in some barley, rye, wheat, and forage grass germplasm and in a few wild species related to wheat. Credit: U.S. Department of Agriculture, Agricultural Research Service.

## Colleges, Universities, and the National System

Colleges and universities have made much use of germplasm for breeding and basic research with the aid of public sector funds. Although the private sector now plays a major role in producing modern varieties of such food crops as corn, sorghum, and soybeans, it is unlikely to assume a larger role in evaluating basic germplasm stocks.

Scientists in public and private colleges and universities have contributed to genetic resources activities, including plant exploration, evaluation, and regeneration. For some investigators, the USDA has provided funds for specific research tasks. Other federal agencies (e.g., NSF) have sponsored the work of university scientists on germplasm evolution, population genetics, taxonomy, and physiology. Examples include the collections of maize assembled in the 1940s by E. Anderson and H. C. Cutler of Washington University and P. C. Manglesdorf of Harvard University; A. F. Blakeslee's Jimson weed (*Datura*) collection; and R. E. Cleland's Evening Primrose (*Oenothera*) collection, which provided benchmark studies in cytology and cytogenetics.

The maize collections of Anderson, Cutler, and Manglesdorf were built on the remnants of an earlier collection developed by G. N. Collins and J. H. Kempton of the USDA and then left to languish. Similar examples can be cited for tomatoes, barley, peas, and their wild relatives. The NPGS has recently begun to assist in developing and maintaining several of these old, long-used collections.

Recently the NPGS has been able to support non-USDA curators of selected germplasm collections. These arrangements should reduce the probability of losing such valuable resources. They allow collections to be managed by researchers with specific knowledge of their contents who can use specialized techniques that may be necessary for maintenance. However, it is important to survey and inventory many other collections, before they are lost. The NPGS is attempting to do this (H. L. Shands, U.S. Department of Agriculture, personal communication, September 1989).

In general, the record of accomplishments of university scientists is excellent. Many of their collections were incorporated into USDA

---

### SORGHUM
PI 264453
*Sorghum bicolor*
Moench

GRIN Data

Origin: Spain
Acquisition: Spain
PI assigned: 1960
Maintenance site:
    Southern Regional
    Plant Introduction
    Station

*Greenbug damage can be compared on sorghum hybrids (left) with and (right) without genetic resistance to the insect. Credit: A. Bruce Maunder.*

It may fairly be asked why large collections of germplasm should be maintained when an individual accession may be used only rarely, if ever. The value of a collection lies not in the frequency of its use, but in the resource it provides when, and if, it is needed. It is thus not surprising that some accessions may possess genetic traits that remain unknown or unrecognized for many years, but which later prove to be almost invaluable.

In 1980, motivated by the threat of a new outbreak of greenbug (*Schizaphis graminum*) in the U.S. sorghum crop, scientists began to screen sorghum accessions in search of genes for resistance. An outbreak of a

collections, but some remain at risk. Increasing emphasis on molecular biology and genetic engineering at universities and colleges has raised concern about the declining interest in breeding, taxonomy, and evolutionary studies that could lead to abandonment of important collections. Recognition of the national importance of research on genetic resources has led to greater USDA-ARS support of at least some of these programs and scientists. There must be broader opportunities for the NPGS to harness the expertise of scientists in colleges and universities.

## Private Industry and the NPGS

Chemical and pharmaceutical companies have supported plant collection activities to find new plant products. Pyrethrum from species of *Chrysanthemum* and anticancer compounds from *Vinca* species are examples. However, once the effective compound is isolated, identified,

---

mutant (new biotype) greenbug on sorghum in the 1960s and early 1970s had cost U.S. farmers more than $100 million in 1968 alone. Varieties of sorghum with resistance to the more common greenbug were eventually developed, but they were defenseless against the new form of the insect. The new biotype, like the earlier one, migrates from wheat and kills by injecting a poison into the plant's tissue.

Resistance to this new greenbug infestation was needed and was found in PI 264453, a cultivated variety of sorghum that had been introduced to the United States from Cordoba, Spain, in 1960. Its origin was most likely Africa, where all sorghums are thought to have originated. Shortly after introduction, this accession was found to be resistant to other greenbug infestations. It was not until 1980, however, that it was found to be one of the few germplasms tested that had resistance to this new outbreak.

Several commercial sorghum hybrids now possess the genes for greenbug resistance from PI 264453. The United States produced 741 million bushels of sorghum in 1987, worth about $1.2 billion. It is estimated that about 1 million acres of sorghum annually are protected in the United States through this genetic improvement. Without a broad collection to search for greenbug resistance, locating the genes to provide protection would likely have been a much lengthier, more difficult, and uncertain task.

---

*"GRIN Data" for the plant introduction (PI) number above represent information contained in the Germplasm Resources Information Network (GRIN). The narrative was prepared from information supplied by A. Bruce Maunder, DEKALB Plant Genetics.*

and synthetically reproduced, there may be little economic incentive for the companies to maintain the germplasm.

The confectionery, perfume, beverage, and specialty food industries are interested in exotic plants and plant products. On occasion, these industries have funded collections of particular genera or species. Collections produced by these activities, however, generally do not become part of the NPGS.

A few, large U.S. seed companies, by contrast, have often contributed to the national system. These companies have regenerated collections in field plots for the NPGS. By growing materials obtained from many countries in a disease-free environment, the seed produced can be assured to be safe for importation. The multinational research programs of some seed companies have thus been used to meet the needs of U.S. genetic resources conservation and development.

## PLANT GERMPLASM ACTIVITIES OUTSIDE THE NATIONAL SYSTEM

### Botanical Gardens and Arboreta

Botanical gardens and arboreta are primarily intended to meet local needs. Their forms and functions may range from university gardens with academic and public service responsibilities, to privately sponsored gardens that are governed by a board of directors. They may or may not have research capabilities. Although many were centers of plant introduction (Plucknett et al., 1987), they are no longer primary repositories of crop genetic resources. They have become increasingly concerned with conserving wild plant species, particularly those that are rare or endangered. On a global basis this has been encouraged by the International Union for Conservation of Nature and Natural Resources (Bramwell et al., 1987) through its Botanic Gardens Secretariat.

The Center for Plant Conservation (CPC), an association of U.S. botanical gardens, maintains rare and endangered U.S. plant species. Cooperating gardens acquire and maintain plants or seeds of these species in their respective regions. The NPGS helps the CPC by providing back-up storage of seeds for its collections at the western regional station and at the NSSL. The CPC's collections are not, however, part of the NPGS.

### Nongovernmental Activities and Private Collections

Farmers, gardeners, and hobbyists have become important conservators of old, obsolete, or heirloom varieties of vegetables, fruits, and

The diversity of heirloom varieties of winter squash is shown in the collections of the Seed Savers Exchange. Credit: David Cavagnaro.

flowers (Office of Technology Assessment, 1985). Grass-roots groups share information, seeds, and plant materials. For example, the North American Fruit Explorers brings together people interested in conserving antique or heirloom fruit and nut varieties. A more organized effort, the Seed Savers Exchange (SSE), maintains a large collection of its own, forms networks of gardeners around the country, and publishes several books related to conserving heirloom genetic resources (Office of Technology Assessment, 1985). Native Seeds/SEARCH (Southwestern Endangered Aridland Resource Clearing House) in Tucson, Arizona, seeks to integrate cultural and biological resources in an effort to conserve locally adapted crops and wild species through in situ and ex situ efforts. Such groups are not formally part of the NPGS, but they may hold valuable germplasm not in its collections.

The SSE is one of the largest and most active nongovernmental groups that preserve plant genetic resources. It is based on a small farm in rural Iowa and depends on individuals who maintain seeds of numerous obsolete heirloom varieties of vegetables and crops. Like the NPGS, the SSE recognizes the problems of duplication, inconsistencies in nomenclature, contamination, inadvertent crossing between varieties, and record keeping. It has developed a database system, trained curators, and produced a variety of written instructional materials. *The Garden Seed Inventory* (Seed Savers Exchange, 1989a) lists all nonhybrid seeds offered for sale by mail-order seed companies in the United States and Canada. It itemizes cultivars that are decreasing in availability and encourages readers to preserve them in their gardens. SSE also has produced a similar listing for fruit, berry, and nut crops available by mail order in the United States (Seed Savers Exchange, 1989b).

Nongovernmental collections generally differ from those of the national system. The SSE preserves old, obsolete, open-pollinated vegetable varieties not available commercially. Active members will often select materials to assure that the desired type is maintained, rather than to preserve genetic diversity. "Off" types are culled rather than preserved for the unusual or rare genes they may possess. Seed is produced every 1 to 3 years and is thus subject to regular environmental selection pressures each time a seed increase is made. In contrast, NPGS accessions are meant to be preserved without genetic change and are held as the basis for breeding and research activities. Thus NPGS collections generally contain a greater proportion of wild species, landraces, and breeding lines. Regeneration and storage methods are designed to reduce the number of times regeneration is needed. The SSE focuses on preserving the general, visually discernable characteristics of an accession, whereas the NPGS is more concerned with preserving all of its genes.

The Living Historical Farms also holds a variety of germplasm. These collections can be sources of old, obsolete varieties that are no longer commercially available. Recent interests in specialty vegetables has raised awareness of many older varieties and led to their commercial use.

The genetic resources maintained by many of these collections are at risk and are not always readily available for research purposes. They should be part of the national collections. Accuracy of nomenclature, authenticity, uniqueness, the extent of internal duplication, and overlap with existing national collections should be determined and assessed. Limited budgets, facilities, and personnel will constrain this work, but these collections should be encouraged and assisted. The NPGS agreement to provide back-up storage for CPC collections is an example of this cooperative effort.

# 3

# Administration of the National System

The collection and management of plant genetic resources through the National Plant Germplasm System (NPGS) are matters of crucial importance to the nation's economy and food security. The U.S. food supply and the nation's ability to compete in and satisfy world food markets depend on these resources. Consequently, the management and coordination of plant germplasm activities at a national level demand broad participation among public and private agencies. The NPGS, a national endeavor, must transcend state and regional boundaries and authorities.

Despite its national responsibilities, NPGS leadership and management are scattered and difficult to discern. The national program leader for plant germplasm, designated by the Agricultural Research Service (ARS) to oversee germplasm activities, is widely regarded as the leader of the NPGS, but this position has little authority over funding, staffing, administration, and program development. The program leader only makes recommendations with regard to germplasm and its management by sites.

The NPGS was designed to be a cooperative network. The U.S. Department of Agriculture (USDA) is the lead agency. Other federal agencies, all of the state agricultural experiment stations, and some private companies and groups participate in the system. A number of advisory groups also guide its activities.

The need to provide a clear framework for the national system has been recognized for several years. A previous report (Office of Technology Assessment, 1987) noted that funding, advice, and administration in the NPGS flow along separate and independent lines of authority. An analysis by the USDA revealed the frustration within that agency over management authority (U.S. Department of Agriculture, 1981:II-16).

The level at which decisions are made is often so obfuscated that accountability is lost. This weakness is not unique to the NPGS, but it is especially critical because of the breadth of the program.

While attempts to address this concern have been made, the administrative weaknesses and clouding of accountability remain. This chapter examines the administrative, organizational, and advisory components of the national system.

## THE NPGS IN THE U.S. DEPARTMENT OF AGRICULTURE

Management of the NPGS network is divided among several agencies within the USDA and state agricultural experiment stations. The ARS holds the largest share of responsibility for management. The diffuse nature of the NPGS with its many sites and cooperators often means that those outside the system do not understand how it functions. Facilities at several locations and a range of expertise and environments are necessary because of the variety of germplasm held.

Although federal NPGS sites have agreed goals and priorities in managing germplasm, conflicts may occur where state or local objectives and plans for funding are different. The constraints these conflicts put on a national program have also been criticized (Office of Technology Assessment, 1981:45) in other USDA research efforts.

The highly decentralized nature of the USDA research system, a source of friction through much of the 20th century, now seems to be accepted and even favored by the States. . . . This dispersion, in fact, has led to criticism that many USDA employees essentially function as State employees and that this in turn has led to a loss of focus on national issues.

Management of the various units of the NPGS is largely the responsibility of the ARS, which operates through an area management structure. The USDA Cooperative State Research Service (CSRS), through its regional structure, supports projects at some NPGS sites. Germplasm conservation, however, requires a national focus. The four regional plant introduction stations, for example, are administered by four separate area directors who support and foster ARS activities important to their respective areas. This decentralized organization is an obstacle to the national coordination and focus crucial to NPGS efforts. Where both CSRS and ARS provide support (e.g., the regional stations), each provides separate administration of its funds. Thus, no single USDA office holds complete authority for the budget and program of the NPGS.

## The Agricultural Research Service

From 1901 to 1953, USDA germplasm exploration and collection were the responsibilities of the Bureau of Plant Industry. In 1953 the bureau, the other scientific bureaus, and the Office of Experiment Stations were combined into the Agricultural Research Service (Office of Technology Assessment, 1981). Germplasm work then became the responsibility of the New Crops Research Branch and, for some major commodities, of various field crop leaders or chiefs. In general, the ARS branch system, as this organization was informally called, centralized finance and decision making and placed line authority largely in the hands of branch chiefs as national leaders.

Several reorganizations since 1953 have altered the balance of financial and decision-making authority within ARS and have seriously affected germplasm activities. The 1972 ARS reorganization switched the agency from the branch system to decentralized management. This shift disrupted the national focus for genetic resources. Line authority was delegated to the western, north central, northeast, and southern regions, each under a regional deputy administrator, and to 22 areas with their own management offices. Over the ensuing years, the regional offices were eliminated and the areas were reduced to eight. The present ARS National Program Staff has little of the authority for budgeting, staff selection, or decision making exercised by the former branch chiefs. Their responsibilities are described as programmatic (i.e., program planning), which is to say, advisory.

ARS activities are managed through a decentralized system of area offices. This system, while responsive to local needs, can hamper a nationally focused program. National coordination of hard red winter wheat research in 1981, which required the concurrence of the deputy administrators for three regions and the cooperation of seven area directors and 11 experiment station directors, is illustrative of the difficulties that can confront a national program (Office of Technology Assessment, 1981). Since that time, regional offices have been eliminated and the number of areas reduced, but decentralized management and the need for multiple concurrences remain.

The NPGS emerged in 1974 as a reorganized national program for germplasm in the United States. The germplasm activities at the regional stations and the National Seed Storage Laboratory (NSSL) were placed

90

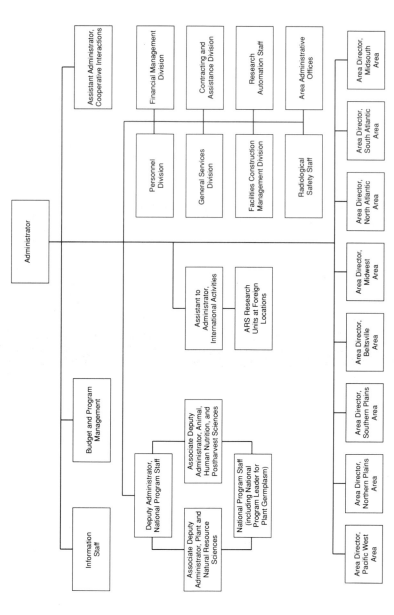

FIGURE 3-1 Illustration of the administrative structure and lines of authority in the Agricultural Research Service showing the position of the national program leader for germplasm within the National Program Staff. (Adapted from a chart prepared by the U.S. Department of Agriculture, Agricultural Research Service, Personnel Division, June 10, 1987.).

under new lines of ARS authority. From 1974 to 1984 the newly created NPGS was managed by a special coordinator, who was the assistant to the ARS deputy administrator of the Plant Germplasm Program. This allowed somewhat greater influence over germplasm activities than before, but authority for implementation remained in the area offices. However, there was better oversight over decisions affecting the national system because the special coordinator reported directly to the ARS deputy administrator.

In 1984 the present relationship between the NPGS and the National Program Staff was established. The special coordinator position was eliminated, and recommendations for germplasm activities became the responsibility of a national program leader for plant germplasm, who is part of the National Program Staff (Figure 3-1). The national line authority for program direction and budget, formerly held by the special coordinator, has been drastically eroded. The program leader now chairs a Germplasm Matrix Team (GMT) comprised of other National Program Staff members. Only with concurrence from the GMT and the cooperation of the area directors can the program leader make recommendations on appropriating funds, hiring staff, and coordinating activities. The program leader's success in executing a national germplasm program depends largely on his or her ability to persuade or to cooperate with the eight area directors and the GMT.

The GMT reviews ARS germplasm activities, offers recommendations and proposals for exploration and evaluation to the ARS deputy administrator, and discusses policy questions. Its members have different primary interests and compete for the same funding sources. This management by committee dilutes the influence of the national program leader for plant germplasm as the leader of the NPGS.

The national system lacks the national and international visibility and influence needed to assure the long-term, continuing support it requires (Christensen, 1989). There has been little sense of cohesiveness within the system, although recent efforts, such as the increased reliance on the Plant Germplasm Operations Committee (PGOC) to promote communication and problem solving, have partly addressed this issue. Because management is area based and decentralized, no individual below the administrator of ARS has line authority for national germplasm activities. To execute an ARS plan, the program leader must, at a minimum, secure the concurrence and approval of the GMT, and the cooperation of the deputy administrator, the administrator, and the relevant area directors for whom germplasm may be a small part of their responsibilities.

This is illustrated by the National Germplasm Resources Laboratory

(NGRL), one of 22 laboratories in the Plant Sciences Institute of the Beltsville (Maryland) Area (Stoner, 1988). It includes plant exploration activities, the Plant Introduction Office (PIO), and the Database Management Unit that oversees the Germplasm Resources Information Network (GRIN), and it manages the crop advisory committees. The scientists in this laboratory report to the head of the laboratory who is in turn responsible to the area director, who evaluates activities and allocates resources. The national program leader for plant germplasm is merely an adviser with no delegated authority over the laboratory or its activities.

Decentralized management is in sharp contrast to the national focus of germplasm work. Germplasm held by individual sites may well have regional significance (e.g., maize at the North-Central Regional Plant Introduction Station), but it also benefits agriculture in all of the regions. NPGS collections also have international significance. For example, many of the collections held in the NSSL have been designated by the International Board for Plant Genetic Resources as international base collections (National Research Council, 1988). The challenge to the USDA is to structure the NPGS so that it has national authority and international visibility but still meets the needs and concerns of its users.

ARS germplasm research appropriations are summarized in Table 3-1. These funds represent all of the ARS research and service efforts that can be classed by their respective administrators as being related to germplasm. While about $26 million to $28 million is spent annually on germplasm work, only about half of that amount is devoted directly to germplasm management at the principal NPGS sites. In fiscal year 1988, $13.8 million was spent at these sites (Table 3-2).

TABLE 3-1   Research Appropriations of the Agricultural Research Service for Plant Germplasm Activities, Fiscal Years 1986–1989

| Activity | Fiscal Year ($000) | | | |
|---|---|---|---|---|
| | 1986 | 1987 | 1988 | 1989 |
| Acquisition | 2,267 | 2,153 | 3,184 | 3,762 |
| Preservation | 6,178 | 6,584 | 9,497 | 10,175 |
| Evaluation | 4,088 | 5,504 | 8,142 | 8,537 |
| Enhancement | 844 | 4,209 | 5,633 | 6,029 |
| Total | 13,377 | 18,450 | 26,456 | 28,503 |

SOURCE: H. L. Shands, U.S. Department of Agriculture, personal communication, September 1989.

TABLE 3-2   Germplasm Research Appropriations of the Agricultural Research Service (ARS) Compared with Responses to a Survey of Funding for Germplasm Activities at the Principal National Plant Germplasm System (NPGS) Sites, Fiscal Year 1988

| ARS Location | NPGS Site Surveyed | ARS Appropriations (dollars) | Survey Response (dollars) ARS | CSRS[a] |
|---|---|---|---|---|
| Arkansas | | | | |
|   Stuttgart | | 177,400 | | |
| Arizona | | | | |
|   Phoenix | | 70,300 | | |
|   Tucson | | 73,700 | | |
| California | | | | |
|   Brawley | National clonal germplasm repository | 40,700 | 36,000 | |
|   Davis | National clonal germplasm repository | 447,400 | 327,000 | |
|   Fresno | | 152,000 | | |
|   Riverside | National clonal germplasm repository | 70,900 | 120,000 | |
|   Salinas | | 89,700 | | |
| Colorado | | | | |
|   Fort Collins | National Seed Storage Laboratory | 2,231,800 | 2,004,000 | |
| District of Columbia | National Arboretum | 616,000 | 343,800 | |
| Florida | | | | |
|   Miami | National clonal germplasm repository | 460,500 | 526,760 | |
|   Orlando | National clonal germplasm repository | 379,100 | 120,000 | |
| Georgia | | | | |
|   Byron | | 47,200 | | |
|   Griffin[b] | Regional plant introduction station | 1,306,300 | 1,439,460 | 175,698 |
|   Tifton | | 104,000 | | |
| Hawaii | | | | |
|   Honolulu | National clonal germplasm repository | 167,900 | 150,000 | |

(*continued*)

TABLE 3-2 *(Continued)*

| ARS Location | NPGS Site Surveyed | ARS Appropriations (dollars) | Survey Response (dollars) | |
|---|---|---|---|---|
| | | | ARS | CSRS[a] |
| Iowa | | | | |
| Ames/Ankeny | Regional plant introduction station | 1,773,500 | 1,307,328 | 353,120 |
| Idaho | | | | |
| Aberdeen | | 765,300 | | |
| Illinois | | | | |
| Urbana | Long-season soybean collection | 529,700 | 273,955 | |
| Indiana | | | | |
| West Lafayette | | 279,900 | | |
| Kansas | | | | |
| Manhattan | | 90,300 | | |
| Maryland | | | | |
| Beltsville | Plant Genetics and Germplasm Institute[c] | 4,199,100 | 2,344,848[d] | |
| Glenn Dale | National Plant Germplasm Quarantine Laboratory (NPGQL) | 903,200 | 812,900 | |
| Minnesota | | | | |
| St. Paul | | 454,800 | | |
| Mississippi | | | | |
| Mississippi State | | 441,200 | | |
| Stoneville | Short-season soybean collection | 748,100 | 225,178 | |
| Missouri | | | | |
| Columbia | | 152,900 | | |
| Montana | | | | |
| Bozeman | | 29,800 | | |
| Nebraska | | | | |
| Lincoln | | 249,000 | | |

TABLE 3-2   (*Continued*)

| ARS Location | NPGS Site Surveyed | ARS Appropriations (dollars) | Survey Response (dollars) ARS | CSRS[a] |
|---|---|---|---|---|
| North Carolina | | | | |
| Oxford | | 45,900 | | |
| Raleigh | | 490,400 | | |
| North Dakota | | | | |
| Fargo | | 623,400 | | |
| New York | | | | |
| Geneva | Regional plant introduction station | 1,256,200 | 825,103 | 131,500 |
| | National clonal germplasm repository | 301,472 | | |
| Oklahoma | | | | |
| Stillwater | | 291,800 | | |
| Oregon | | | | |
| Corvallis | National clonal germplasm repository | 950,600 | 743,425 | |
| Pennsylvania | | | | |
| University Park | | 96,900 | | |
| South Carolina | | | | |
| Charleston | | 405,900 | | |
| Florence | | 44,700 | | |
| Texas | | | | |
| Brownwood | National clonal germplasm repository | 59,100 | 103,188 | |
| Bushland | | 128,200 | | |
| College Station | Cotton collection | 826,600 | 170,989 | |
| Temple | | 75,000 | | |
| Utah | | | | |
| Logan | | 511,900 | | |
| Washington | | | | |
| Prosser | | 286,100 | | |
| Pullman | Regional plant introduction station | 1,222,100 | 1,270,866 | 245,270 |
| Wisconsin | | | | |
| Madison | Interregional Research Project-1 (IR-1) | 453,500 | 76,400 | 132,251 |

(*continued*)

TABLE 3-2    (Continued)

| ARS Location | NPGS Site Surveyed | ARS Appropriations (dollars) | Survey Response (dollars) ARS | CSRS[a] |
|---|---|---|---|---|
| Puerto Rico | | | | |
|    Mayaguez | National clonal germplasm repository | 343,400 | 237,283 | |
| Virgin Islands | | | | |
|    St. Croix | | 236,300 | | |
| Headquarters | | | | |
|    National Program Staff | | 1,056,300 | | |
| Total | | 26,456,000 | 13,759,955 | 1,037,839 |

NOTE: The survey includes only those funds that are provided by the ARS and the Cooperative State Research Service (CSRS) of the U.S. Department of Agriculture. The in-kind contributions of state agricultural experiment stations and local universities, such as laboratory space, equipment, and other support, were not included. Most respondents reported "net to location" amounts, which do not include an overhead for administrative costs of 10 percent that is deducted from ARS appropriations.

[a] CSRS provides funding only to selected sites.

[b] Formerly referred to as Experiment, Georgia.

[c] In 1988 the Plant Genetics and Germplasm Institute, which included the Germplasm Services Laboratory (GSL), was renamed the Plant Sciences Institute of the Beltsville (Maryland) Area. The activities of the GSL included the National Small Grains Collection (NSGC), the Plant Introduction Office, administration of NPGS-sponsored plant exploration activities, the Germplasm Resources Information Network, and coordination of crop advisory committee activities. In 1989, administrative oversight of the NSGC was removed from the GSL and the facility was relocated to Aberdeen, Idaho. In 1990, the GSL became the National Germplasm Resources Laboratory, and it became responsible for administration of the NPGQL.

[d] The budget figure is the net to location amount for the GSL as reported in Stoner, A. K. 1988. Program review. Germplasm Services Laboratory. Plant Sciences Institute, Agricultural Research Service, Beltsville, Maryland, November 14 (photocopy).

Also included in the ARS appropriations are costs for activities at other locations, which are identified by their national program leaders, research leaders, or others as germplasm related. Many of these efforts may be more closely allied to activities other than those of germplasm management (e.g., breeding activities). The ARS appropriation is not, therefore, a budget for the National Plant Germplasm System. Further, the many competing interests for these funds have constrained their allocation in a way that is responsive to the needs of the system.

## *System of Personnel Classification and Promotion*

Federal research and service scientists within the NPGS are governed by the classification and promotion system of the ARS. The system consists of four categories. The term *category* is a designation for a group of personnel positions having similar characteristics (U.S. Department of Agriculture, 1986). Some positions may include duties that apply to more than one category. The ARS categories that have been established for professional scientific positions are research scientist (category 1) and service scientist (category 4). Category 2 (research affiliate) pertains to a short-term position that is intended primarily for postdoctoral study. Category 3 (support scientist) positions provide assistance and support to the research efforts of category 1 or 4 scientists.

The position of research scientist pertains to an employee whose highest level of work, for a major portion of the time, involves personal conduct, or conduct and leadership, of theoretical and experimental investigations primarily of a basic or applied nature. For example, such work would involve determining the nature, magnitude, and interrelationships of physical, biological, and physiological phenomena and processes; or creating or developing principles, criteria, methods, and a body of knowledge generally applicable for use by others.

The position of service scientist pertains to an employee who serves as a project or program leader for, or who personally performs, work involving professional scientific services to the public or to other governmental agencies. These services include identification of animals, plants, or insects; diagnosis of diseases; mass production of plants, animals, or insects; collection, introduction, and maintenance of germplasm or specimens; vaccine production; education, extension, or technology transfer activities; and nutrient data and food intake surveys.

The Research Position Evaluation System is a personnel assessment mechanism used to evaluate category 1 (research) scientists. It is based on what is termed the "person-in-the-job" concept. Category 1 scientists have open-ended advancement potential based largely on research accomplishments. By building a research program and producing the evidence of a publication record, category 1 scientists can receive regular promotions. In the NPGS, however, category 1 scientists can encounter difficulty in gaining promotion because their duties, such as seed regeneration, evaluation, or oversight, do not necessarily lead to published research papers.

Category 4 scientists are not covered by the Research Position Evaluation System. Instead, they are evaluated by ARS position classification specialists who use classification standards based on the administrative complexities of the job and the number of people supervised. Thus category 4 scientists can be limited in advancement by their job classification. Such limitations on service-oriented career advancement do not exist in parallel agencies, such as the Soil Conservation Service or the Forest Service. Category 4 positions are seen by NPGS scientists as professionally limiting, although they permit more freedom to address germplasm activities.

*Risien's pecan tree still grows on the banks of the San Saba River, Texas. Credit: Tommy E. Thompson.*

## PECAN
PI 518116
*Carya illinoinensis*
(Wangenh.) K. Koch

GRIN Data

Origin: Texas, United States of America
Acquisition: Texas, United States of America
Common name: Pecan
NPGS received: December 21, 1987
Year PI assigned: 1988
Life form: Perennial
Form received: Cuttings
Improvement status: Wild
Local names: San Saba, Eggshell (Texas), Papershell (Texas), Paper Shell, Risien, Risien's Paper Shell, Royal

Although its plant introduction number was assigned in 1988, PI 518116 has a history that dates back to the nineteenth century, with discovery by Edmond Risien, a trained cabinetmaker whose passion in life was the pecan. Risien came to the United States from England in 1872 and 2 years later settled in the central Texas town of San Saba on the banks of the San Saba River. At that time the town was a busy market center where wagonloads of buffalo meat, venison, and pecans were sold. Risien became interested in this native American nut, which grew throughout the region. He offered a $5

## The Cooperative State Research Service

The Hatch Act (Public Law 84-352) mandates funding for regional research projects carried out jointly by regional plant introduction stations and state agricultural experiment stations. Funds from CSRS are allocated annually in accordance with the recommendations of the CSRS Committee of Nine (experiment station directors) and the respective regional associations of experiment station directors. The funds decline when overall appropriations are reduced. This regional line of authority and funding is largely independent of the ARS.

---

prize to the person who could bring him a sample of the best pecan nut.

Upon awarding his prize, Risien asked to see the tree from which the prizewinning nuts had come. To his horror, the owner had cut all but one limb from the tree in an effort to harvest all of its nuts. Risien was so taken with the quality of the pecans from this tree that he ultimately bought the land on which it grew as well as the surrounding 314 acres. The tree itself was situated on the banks of the San Saba River, at its confluence with the Colorado River. He named the tree, San Saba, and intended to develop a pecan orchard with nuts that had the shape, color, and thin shell he prized.

He soon discovered that trees grown from San Saba seed varied tremendously in their size, shape, and growth characteristics. Nuts from these trees ripened at different times and were equally variable in shape, size, flavor, color, and shell thickness, with very few (reportedly only 2 of 1,000) bearing nuts of the sort he sought. Undaunted, Risien began a breeding, selection, and improvement effort that was to last until his death in 1940. The parent stock he used was the tree he had purchased or the trees grown from its cuttings or grafts. By crossing the tree with others and selecting among the offspring, he developed many popular varieties. One, Western Schley, occupies more than 75,000 acres of today's pecan orchards.

About one-sixth of all grafted and budded pecan trees in the United States are descended directly from Risien's original tree. Cuttings from that tree have been propagated at the Brownwood repository and samples from them are used to trace parentage in pecans and to study the heritability of desirable traits.

---

*"GRIN Data" for the plant introduction (PI) number above represent information contained in the Germplasm Resources Information Network (GRIN). The narrative was prepared from information supplied by Larry J. Grauke, National Clonal Germplasm Repository, Brownwood, Texas, and Tommy E. Thompson, Agricultural Research Service, Brownwood.*

The funding of the regional stations through the CSRS is another example of decentralized management and the need for coordination. According to a survey (Table 3-2), CSRS provided $900,000 to the four regional stations and the Interregional Research Project-1 (potatoes) in fiscal year 1988. The ARS national program leader for plant germplasm has no authority over such funds and is only the advisory representative of a separate service. The CSRS is not represented on the GMT and does not presently have a staff position delegated to address plant genetic resources programs.

## State Agricultural Experiment Stations

The state agricultural experiment stations are associated with the land-grant universities of the United States. They are linked in research reporting and federal funding through the CSRS, as noted above. Federal funding of the experiment stations is typically about 20 percent of their annual budgets. The remainder is derived from state government appropriations, gifts, contracts, and grants from individuals, organizations, and state crop commodity boards. The experiment stations carry out extensive basic and applied research programs. A large effort is dedicated to plant genetics, breeding, and germplasm enhancement. In 1986, for example, more than 2,600 research projects related to genetics and crop improvement were reported through CSRS.

The experiment stations are sources of new varieties of many crops and contribute improved germplasm to private enterprise. They develop, evaluate, and introduce new crops and new products from existing crops for public use. The stations are major users of genetic resources. Based on data provided by CSRS, experiment station work on germplasm, of which nearly 70 percent was devoted to genetic analysis, enhancement, and variety development, totaled about $155 million in 1986 (C. O. Qualset and L. W. Gallagher, University of California at Davis, personal communication, June 1989).

Collections at experiment stations include breeding lines, genetic stocks, landraces, and wild species related to cultivated crops. However, the degree to which the stations' collections duplicate those of the NPGS is not known. Some collections (see Table 2-5) are commonly considered to be part of the national system, but it is likely that many breeding lines and genetic stocks are not in NPGS collections.

The experiment stations that host NPGS facilities have historically provided significant in-kind support. The relationship benefits the NPGS, which has access to facilities, services, and land to maintain its germplasm, while the experiment stations benefit from the increase in local scientific staff supported by ARS through the NPGS. Local contributions commonly include land, buildings, maintenance, computer and other support services, and library privileges.

Unfortunately, the cooperative nature of an NPGS site and an experiment station may at times become strained. The extent of in-kind contributions to NPGS sites, such as laboratory, office, or greenhouse space, depend on budget realities at the stations. When their budgets decline or the competition for funds increases, in-kind contributions may be reduced or eliminated. Without additional funds from either CSRS or ARS, operations at NPGS sites suffer.

Where a site depends on local cooperation and in-kind support for its activities, arrangements should be periodically reviewed by the NPGS. Cooperative agreements should clearly state the roles of federal and local cooperators, and they should be sufficiently long term as to be unaffected by changes in local priorities or needs.

Collaborations are also adversely affected by delays in the implementation of cooperative agreements and in the annual process for congressional and administration approval of the federal budget. When passage of the budget is late, the receipt of funds by state scientists for cooperative research activities can be delayed by months beyond the appropriate time for critical germplasm operations, such as planting, thus preventing the purchase of needed supplies or the hire of necessary labor.

### Other USDA Cooperators

The NPGS cooperates with USDA's Animal and Plant Health Inspection Service (APHIS) to operate the National Plant Germplasm Quarantine Center in Beltsville, Maryland. The Soil Conservation Service (SCS) of USDA evaluates NPGS accessions for their various soil conservation programs and may assist in some regeneration efforts. Representatives of APHIS and SCS may participate on various advisory committees in the NPGS, but these agencies provide no direct management or funding for the system.

## ADVISERS TO THE SYSTEM

An earlier USDA review (U.S. Department of Agriculture, 1981) criticized the NPGS for lacking a clear understanding of the composition, responsibilities, guidelines, and limitations of its advisers. Some attempts have been made to reduce overlaps and conflict, but there is still no clear mechanism in the NPGS for using the advice of its many committees. The composition and responsibilities of the principal advisory bodies are summarized below.

## The National Plant Genetic Resources Board

The National Plant Genetic Resources Board (NPGRB) was established by the secretary of agriculture in 1975, in part because of the concerns expressed in the wake of both the 1970 corn blight and the release of the report, *Genetic Vulnerability of Major Crops* (National Research Council, 1972). The NPGRB meets at least twice a year and advises the secretary of agriculture and the officers of the National Association of State Universities and Land-Grant Colleges (National Plant Genetic Resources Board, 1984) on national policy related to the problems, needs, and welfare of the nation's plant genetic resources activities as they affect the food production system. It also establishes priorities for safeguarding plant genetic resources. Members are scientists and administrators from the public and private sectors who are appointed by the secretary of agriculture to serve a 2-year term. They may be reappointed for two more consecutive terms.

The board is chaired by the USDA assistant secretary for science and education; a vice-chair is appointed from its membership. The assistant secretary's office provides an executive secretary and financial and personnel support. As a federal advisory committee, the board must be rechartered periodically by the secretary of agriculture.

## The National Plant Germplasm Committee

The National Plant Germplasm Committee (NPGC) emerged in 1974 from the previous New Crops Coordinating Committee, by mutual agreement of the ARS and the state agricultural experiment stations (Council for Agricultural Science and Technology, 1984; White et al., 1989), to represent the user community. It was intended to facilitate coordination among agencies and to be a source of information for administrators and program leaders working in or with the NPGS. Its members include scientists and administrators from the ARS, CSRS, experiment stations, and the private sector.

According to its charter, the NPGC meets at least once a year, and its functions are the following (Jones and Gillette, 1982):

● Coordinate the research and service efforts of federal, state, and industry units engaged in the introduction, preservation, evaluation, and distribution of plant germplasm, through representation of all of the units' views by committee members.

• Develop policies for the conduct of the national plant germplasm program and for its relationships to international plant germplasm programs.

• Develop research and service proposals and justification for adequate funding of regional and national plant germplasm activities.

• Advocate mutually agreed upon proposals with experiment station associations and USDA agencies.

• Serve as the principal way in which station interests can be presented and harmonized with federal interests at a technically informed level.

### The Crop Advisory Committees

Crop advisory committees are crop-specific groups that provide the NPGS with expert advice on germplasm collection, management, exploration, crop descriptors, evaluation, and enhancement. On the committees, crop specialists include breeders, geneticists, pathologists, and entomologists who are considered the best qualified to assess the status of collections, vulnerability, improvement efforts nationally, foreign scientific developments, the impact of new technology, and how well users' needs are met (National Plant Genetic Resources Board, 1984, Shands et al., 1989; White et al., 1989). Thirty-nine committees have been established. They are intended to review research plans; report on national and international developments; make recommendations on germplasm exploration, evaluation, and enhancement and on training, staffing, and facilities; and provide a forum for commodity groups to make their concerns known to the NPGS.

The committees produce reports, analyses, and recommendations. Their reports vary in detail, accuracy, and comprehensiveness. They are received, compiled, and filed by the NPGS without organized review, analysis, or response. The reports are not widely disseminated or published in the scientific community, although very brief summaries appear in *DIVERSITY*, an international news journal for the plant genetic resources community published by Genetic Resources Communications Systems, Inc. The committee reports are not ignored, but there is no mechanism for using them to set national priorities and develop plans.

The committees are administratively supported, in part, through the National Germplasm Resources Laboratory, but receive no travel or other operational support apart from an annual meeting of the chairs, curators, and other interested individuals funded by ARS. The committee members are volunteers who must either obtain funds to travel to meetings from other sources or pay for them personally. To facilitate attendance, meetings are often held in conjunction with other scientific or professional meetings. This approach, however, does not ensure the participation of all members.

### Technical Committees and Technical Advisory Committees

Each regional plant introduction station receiving CSRS research funds has a regional technical committee (TC). An experiment station director serves as its regional administrative adviser, and its members include a representative from each experiment station in the region and representatives from participating USDA agencies. Other cooperating agencies, such as the Bureau of Land Management of the U.S. Department of the Interior, may also be represented.

There are interregional administrative advisory committees for the two interregional research projects and an interregional committee that functions as a TC for each interregional project. These committees meet separately and function independently of other advisory groups concerned with the site. Technical advisory committees (TACs) provide advice on technical and scientific issues to clonal repositories. They do not exercise authority over programs at the site, but are a source of expert advice. A TAC is composed of individuals chosen for their scientific and technical expertise. For some sites a crop advisory committee may fulfill all or part of the TAC's role.

The TCs and TACs are advisory to specific sites and independent of other advisory groups. TCs are responsible to CSRS, and are not under the purview of the national program leader for plant germplasm or any germplasm advisory group of the ARS. Thus, there is frequently a lack of coordination and leadership among advisory groups regarding national program requirements.

### The Plant Germplasm Operations Committee

The PGOC includes the curators of the major collections, selected ARS research leaders, and the lead people from ARS-NPGS support offices (e.g., the National Germplasm Resources Laboratory). Its other members include representatives from the regional stations, repositories, NSSL, the National Small Grains Collection, and the cotton, soybean, and potato collections. The chair is elected by the committee from its membership. This committee has been effective in maintaining communication and promoting cooperation among NPGS sites. However, it has no administrative authority. The PGOC is an ARS group that translates administrative decisions into action.

The PGOC discusses policy only as it relates to germplasm operations. Its members are involved in site management. The committee plans NPGS activities and its participants have a sense of cohesiveness. This is the only NPGS activity where the program leader exerts a strong degree of leadership and can directly affect NPGS activities.

# 4

# Prescription for Effectiveness

The United States, through the National Plant Germplasm System (NPGS), distributes free of charge more germplasm around the world than any other nation and plays an important role in efforts to manage and protect the world's crop genetic resources. The creation of the national system more than 15 years ago was intended to herald an efficient effort aimed at coordinating activities throughout the country. Instead the system has been burdened by a cumbersome administrative structure inappropriate to managing a program that has substantial national and international responsibilities. A plethora of advisory and administrative bodies make it difficult to discern where, or if indeed, there is any central germplasm leadership and authority in the United States. Central, unified, budgetary authority for NPGS activities is similarly lacking. These dispersed, sometimes overlapping administrative, advisory, and budgetary components have often confused and hampered the effectiveness of NPGS.

*The administrative and advisory organization of the National Plant Germplasm System should be structured to provide for efficient national coordination.*

The administrative structure of the NPGS is inefficient and far too complex. The great strategic importance of plant genetic resources requires that the system be administered centrally, at the national level. Stronger and clearer liaison among the cooperating units and agencies is needed if the NPGS is to address effectively the many issues that confront it. More direct lines of authority must be vested in a central management unit to address long-standing needs and concerns and to reduce the complex bureaucracy separating individual site activities from those who should exercise coordinated national management. To be effective, however, this national authority must possess the capability of linking program and policy development with budget authority. Placing greater decision-making and budgetary authority in a central unit and reducing the administrative inputs will reduce the multiple authorities to which individual sites are responsible and will enable the NPGS to deal directly with national needs.

Furthermore, a centralized system would provide much needed coordination, guidance, and direction to U.S. policies regarding the collection, exchange, and use of genetic resources around the world. A centralized NPGS could act as the liaison to other parts of the U.S. Department of Agriculture (USDA), other executive branch departments (e.g., the U.S. Department of State), Congress, industry, and other private efforts to manage germplasm. In so doing, the many and sometimes disparate interests and concerns of these groups would receive greater attention when policies, directions, and budgets for the NPGS are developed.

## ACHIEVING A NATIONALLY MANAGED SYSTEM

The primary barriers to consolidated, central management of the NPGS are that it is a dispersed system and that clear authority and responsibility for program direction and budget are not vested in a single office or individual. There is no distinct budget for the NPGS. Because responsibility for activities and budgets are dispersed, there is no well-defined mechanism for assuring that budgets accurately reflect or address the needs of the system. Its support is derived as a portion of the funds more broadly directed toward germplasm-related work.

Because the Agricultural Research Service (ARS) is the primary source of funds for most of the principal NPGS sites, the NPGS is frequently perceived as an ARS responsibility. Other public and private entities, however, play important roles. This perception of ARS responsibility has sometimes hampered interagency cooperation. Within ARS, management of germplasm activities through the Germplasm Matrix Team and the ARS area directors has been an obstacle to achieving a coordinated nationally focused program.

More direct control must be vested in a central, national authority. The policies and directions of the NPGS should originate from this authority and be overseen by a national board that is representative of the wide array of agencies, offices, and public and private groups that work with or are served by the NPGS. The office must provide liaison with agencies, offices, and groups at the national and international level with regard to U.S. germplasm activities, and, where appropriate, have authority to foster cooperative activities. The committee has identified two options to give the NPGS greater visibility within the USDA and to simplify and centralize its management: the creation of a reorganized national system apart from the ARS, or the elevation of the NPGS within the ARS.

## Creation of a Reorganized System Outside the ARS

The NPGS could be removed from the ARS to become a separate entity within the USDA's Office of Science and Education. It would cease to be the responsibility of the National Program Staff, and would be overseen by an administrative unit reporting to the assistant secretary for science and education. The unit would have direct responsibility for NPGS budgets, staffing, and program execution. Sites and program activities would be administered directly by this new body rather than through the ARS areas or the regions of the Cooperative State Research Service (CSRS). Cooperative support from ARS, CSRS, or others would be provided for specific activities, but the national office would coordinate activities and funds. The National Plant Genetic Resources Board (NPGRB) would provide oversight and guidance for policies and programs.

The reorganized system would administer sites, collections, international activities, germplasm acquisition, data and germplasm management, research, and advisory and other activities related to managing plant germplasm in the United States. The new NPGS should also, with guidance from the NPGRB and through appropriate government offices (e.g., U.S. Department of State), provide liaison for bilateral cooperative agreements and for international germplasm activities with, for example, the Consultative Group on International Agricultural Research (CGIAR) centers and the Food and Agriculture Organization (FAO) of the United Nations.

Removing the administration of the national system from the ARS would provide more direct line authority and budgetary control from the system leader to individual sites. Budgets could be administered centrally and activities coordinated nationally. ARS scientists or others with responsibilities in addition to germplasm could hold joint appointments. In this way, salary and other costs attributable to germplasm could be covered by the NPGS. Given a clear national plan for managing germplasm in the United States, the NPGS would be more responsive to the needs of individual sites and to the requirements of international collaboration.

Moving the NPGS out of the ARS would pose some difficulties. Such a change could distance germplasm management from basic research efforts that have proved to be important parts of the overall activity. An NPGS that is organized outside and parallel to the much larger ARS could have reduced visibility when budgets are developed or other resources allocated. As a relatively small unit, it might be difficult for the NPGS to obtain cooperation from larger services, such as the ARS or CSRS.

Such a reorganization would necessitate creation of new administrative staff to provide services related to personnel, contracts, accounting, and purchasing. The NPGS could consider obtaining administrative services through cooperation with either ARS or CSRS, but not without providing funds to accomplish them. Difficulties might also arise when the NPGS develops policies and procedures that depart from those in their cooperating agencies. Providing these services within an independent NPGS would be possible, but it would require additional funds to achieve, beyond those presently allotted to germplasm work in ARS.

The association of the national system with the basic and mission-oriented research of the ARS has been an asset, particularly for the application of new technologies to germplasm management. Development of methods for the cryopreservation of seeds and tissues by ARS researchers at the National Seed Storage Laboratory (NSSL) should, for example, lead to improved methods for maintaining materials in long-term storage. Basic research at that same facility on the biophysics of water in dried seed could lead to new technologies for storage and viability assessment. ARS researchers involved in germplasm enhancement or evaluation are significant users of NPGS germplasm, whether located at an NPGS site or elsewhere. Removing the national system from ARS could weaken the important link between germplasm management and basic research.

As in many organizational structures, power bases can be very important. The power base for an independent NPGS would be small. The NPGS could find itself competing, rather than cooperating, with the ARS for funds, staff, and equipment. Where sites would be occupied by both NPGS and ARS scientists, or those with joint appointments in the two units, competition for oversight of resources and facilities could exist. There are, of course, conflicts similar to this now. The question is whether they could be better resolved outside rather than inside the ARS.

### Elevation of the NPGS Within ARS

It may be possible to continue the administration of the NPGS within the ARS. However, the responsibility for oversight of NPGS activities would have to be elevated from the National Program Staff and vested in a director who would report to the administrator of ARS on budgets and program direction.

This change would require an organization that would, to some degree, push aside the present ARS system of area administration. Responsibility for budget, staffing, and program direction would rest with the central office. This change would also obviate the need for the Germplasm Matrix Team, although recommendations on programs could be sought from the National Program Staff as needed to coordinate activities.

The NPGS director should be responsible for formulating an overall budget for the system that would take account of funding and other resources provided by other cooperating agencies (e.g., CSRS, state agricultural experimental stations). In developing a budget, the director should address the concerns and priorities of the national system as identified by the NPGRB. The NPGS budget should be a separate element of the ARS budget, clearly distinguished from other ARS activities. An annual report should be made to the NPGRB by the leader of the NPGS on the effectiveness with which the board's recommendations for budget and program were addressed.

The USDA must develop an NPGS organization with minimal bureaucratic and administrative entanglements and maximal independence. Of the two options, the committee favors creation of a reorganized NPGS outside ARS as the most likely to bring about the positive administrative and advisory changes that it recommends. The second option would perpetuate many of the current administrative constraints on the operation of the NPGS.

Addressing the crucial needs of the NPGS will require significant changes in budget responsibility and in the way the NPGS is organized and managed. The committee cautions the USDA not to respond to these recommendations solely by generating more cooperative or informal agreements. Such agreements are valuable mechanisms for enabling the NPGS to achieve important specific goals, through the sharing of resources and responsibilities among NPGS cooperators. However, they cannot alone provide the sharp national focus and central authority needed.

## CHANGES IN THE ADVISORY STRUCTURE

At present, the responsibilities of advisory units within the NPGS often overlap. In many cases, no mechanism exists for considering the advice or reports of these groups. The responsibilities of the advisory groups must be clearly defined and their advice must be used in developing the system's activities, programs, and budgets. Site-specific advisory committees (such as the technical advisory committees described in Chapter 3) could be established by individual sites in cooperation with the central management office. These committees should provide expert technical and scientific advice in support of the site's nationally mandated activities.

*The National Plant Genetic Resources Board must have greater independence as an adviser on national and international policies.*

During its first years the NPGRB showed strong leadership in its policy recommendations and monitored their implementation diligently. By the early 1980s its role had diminished. Little regard was paid to the terms of appointment and rotation of membership, and a wholesale turnover of members resulted in a lack of continuity. The executive secretary's position, initially filled by a CSRS senior staff person, was given to a member of the ARS National Program Staff. The style of operation of the board also changed as ARS appeared to influence its agendas, recommendations, and activities more than in the past.

Despite the intent that "the Board will be composed of individuals with diverse capabilities distinguished by their knowledge and interest in plant genetic resources management" (National Plant Genetic Resources Board, 1984), the diversity of representation narrowed. The minutes of board meetings reveal that informational items were discussed and that substantive policy issues, such as plant patenting and international activities, were neither discussed nor pursued. The board produced few written formal or official statements, recommendations, or positions. Thus, it had little influence over genetic resources issues.

Although the current chair of the NPGRB has made the board more vocal about and responsive to policy issues, the board must have greater independence in advising on plant genetic resources policy. The focus of the board's activities must be clearly distinct from that of other advisory bodies, such as the National Plant Germplasm Committee

(NPGC) or the Plant Germplasm Operations Committee (PGOC). Its primary concerns should be genetic resources policy, strategy, and international cooperation. Its recommendations should be incorporated into the formulation of NPGS budgets and programs.

The NPGRB should have greater independence from those who receive its reports and advice. At present the assistant secretary for science and education, as chair, both transmits and receives the board's reports and advice for the secretary of agriculture. The board would be better served if its chair were elected from the membership. The chair should be the board's advocate in presenting its decisions and recommendations to the secretary of agriculture and the leader of the national system. The chair, on behalf of the board, should transmit an annual report summarizing the board's activities to the secretary of agriculture, the relevant congressional committees, the National Association of State Universities and Land-Grant Colleges, and other interested parties. The report should address U.S. germplasm activities and the effectiveness of the NPGS in achieving the board's budgetary and programmatic recommendations.

The executive secretary of the NPGRB is responsible for administrative activities. This individual should be independent of the ARS National Program Staff and, with the chair, should aid in developing and setting agendas.

Representation on the NPGRB should include public and private sector scientists and administrators with responsibilities and expertise for managing or using plant germplasm. Representatives from other relevant federal offices, such as the Departments of State and Interior would ensure that the board's deliberations include the concerns of all NPGS participants. Private, nonprofit groups with interests in managing and conserving plant genetic resources should also be represented.

*The National Plant Germplasm Committee should be disbanded.*

The NPGC is superfluous and should be eliminated. The committee was once an effective advocate for the NPGS, but today its members' expertise in and commitment to germplasm varies considerably. Some members, while interested in the subject, have little prior experience with the germplasm system. As a consequence a major part of the NPGC's work has been to educate its members about plant genetic resources. A clear role distinct from other advisory groups no longer exists for the committee. No designated individual or office within the NPGS receives its reports and advice, and its ability to influence policy is limited.

*The crop advisory committees should be provided financial support, and a mechanism should be created to use their reports when developing policies and priorities.*

The crop advisory committees could be valuable in assessing the status and needs of NPGS collections. They need encouragement and financial support, but have received little of either from the National Plant Germplasm System. As a result their impact and effectiveness are reduced.

The committee's reports can be very useful as sources for developing plans and priorities for the NPGS. However, there is no mechanism to ensure they are used to set national priorities and develop plans. The more assertive committee members go directly to the NPGS, the USDA, and even to the Congress, either independently or through a commodity group, to obtain action on their concerns. This can lead to unbalanced treatment for some crops and to priorities that take no account of the relative needs of other crops.

Recently the ARS national program leader for plant germplasm has brought together the chairs of the crop advisory committees, crop curators, and others for annual meetings that have been useful for promoting communication. However, well-attended, regular meetings of the crop advisory committees are needed to discuss tasks and produce reports. Some support to chairs for administrative expenses would facilitate communication with the membership between meetings when urgent questions arise.

These committees must be developed further as key elements of the national system. If they are to receive a minimal level of support, their numbers should be reassessed. There should also be a central review of all of the reports by an existing group, the PGOC, or a committee drawn from the committees' chairs.

*The Plant Germplasm Operations Committee should be given responsibility for advising the leader of the National Plant Germplasm System on management, operations, and priorities.*

The PGOC has become an effective and responsive advocate for the needs and priorities of site managers in the NPGS. It provides a forum for debating the various needs of sites and collections that allows for the balancing of divergent priorities. It should report to the leader of the NPGS on matters pertaining to operations and functions at germplasm sites, and to provide advice on coordinating and developing management plans and priorities derived, in part, from the reports of the crop advisory committees.

## GERMPLASM ACQUISITION AND COLLECTIONS

New germplasm should be acquired in response to a long-range plan based on analyses of present holdings and future needs and goals. As collections grow in size it is increasingly important to develop procedures that allow them to be used easily and managed efficiently. This will require the NPGS to give continuing attention to quarantine, information management, development of criteria for entry of new accessions, and the problems of managing large collections (Chang, 1989).

### Plant Exploration

*The National Plant Germplasm System should develop a comprehensive plan for plant exploration.*

Successful explorations depend, in part, on clear, scientifically based objectives. In the past, lack of a plan for exploration has resulted in some crops receiving greater attention while others, with few champions, went unserved.

Until very recently, the guidelines for plant exploration activities were too rigid and complex for even a relatively simple collecting trip. Flexibility is needed to approve, fund, and expedite the various endeavors in plant exploration. These may involve a single collector, a team of individuals, multinational fieldwork, collection of a precisely located endemic species, or collection of many species over a wide range. Past difficulties have led to a decline in requests for support and for exploration activities, and to criticisms about the lack of these activities in the national system. In recent years, actions have been taken to address these deficiencies.

Since 1988, the national system has begun to develop priorities based on deficiencies in its collections and on expected germplasm needs. The crop advisory committees can play an important part in priority development. Qualified people are being sought to collect germplasm in accordance with established priorities and standards. This approach is a considerable departure from the past practice of assuming that exploration proposals submitted to the Germplasm Matrix Team through the then existent Plant Exploration Office would reflect the appropriate priorities for the NPGS.

It is very important to address national and local concerns when planning explorations. Cooperative efforts that include U.S. and local scientists working together throughout a growing season should be sought through FAO or the International Board for Plant Genetic

Resources (IBPGR), or through bilateral agreements. Local scientists should participate in exploration and collection, and receive samples of all of the accessions obtained.

Finally, germplasm exchanges can be hampered by international trade embargoes that restrict the shipment of agricultural products or commercial grain. Germplasm exchange should be exempt from such embargoes.

### The Plant Introduction Office

*The role of the Plant Introduction Office within the national system should be clearly defined.*

The Plant Introduction Office (PIO) plays an important and highly visible role in international cooperation and exchange. These activities should be centrally managed within the NPGS. In the past, its international visibility has led to the incorrect assumption that the PIO or its leader controlled the NPGS.

The PIO should be the site of germplasm entry and assignment of plant introduction (PI) numbers, and the validation of documentation, nomenclature, and site of origin. Its activities should be clearly within the management jurisdiction of the NPGS leader.

### Quarantine

*The National Plant Germplasm System should continue to seek the development of policies, procedures, and cooperative arrangements that promote the safe, yet rapid and efficient, acquisition of germplasm.*

Quarantine policy, under the regulation of the Animal and Plant Health Inspection Service (APHIS), attempts to prevent the importation of disease and insect pests not indigenous to the United States. Approximately 90 percent of the germplasm that enters the United States moves relatively unimpeded through quarantine, following routine inspection and occasional fumigation, on arrival. Some 8 percent of items entering the country are placed under a postentry quarantine while 2 percent, about 10 genera, mostly vegetatively propagated, are prohibited from entry (H. Waterworth, U.S. Department of Agriculture, personal communication, September 1987). These percentages, however, do not reflect some important materials that because of quarantine restrictions are simply never acquired. In the past, some accessions of *Prunus* (e.g., plums, nectarines, apricots, cherries, peaches) have been delayed in quarantine for more than two decades (S. M. Dietz, U.S. Department of Agriculture, personal communication, July 1990). Because

Accessions of glabrous apricots from Alma-Ata in the south central Soviet Union were introduced into the United States in July 1990 and placed under quarantine. They cannot be made available to researchers until tests to detect plant pathogens in them have been completed, which can take several years. Credit: Calvin Sperling.

of a lack of facilities to grow them under quarantine in the United States, a number of items (such as corn and sorghum accessions from Africa and Asia) must be grown in Europe or Latin America before becoming a part of the 90 percent that pass quickly. An offshore quarantine site, such as that being developed by ARS in St. Croix under a permit from APHIS, will greatly expedite the entry process for many accessions.

Recent agreements between ARS and APHIS have promoted cooperation on importing germplasm for scientific purposes. The National Plant Germplasm Quarantine Center near Beltsville, Maryland, run jointly by NPGS and APHIS, was established to facilitate exchange and importation and to eliminate a rapidly growing backlog of germplasm materials. However, the center's isolation areas, greenhouses, controlled environment rooms, laboratories, and staff members will be insufficient to process the expected volume of materials. ARS and APHIS are seeking funds to expand these facilities.

Cooperation with other sites, such as the national clonal germplasm repositories, may also relieve the burden on APHIS facilities and speed the release of materials. Arrangements can be made to transfer germplasm to selected facilities under APHIS-approved protocols of postentry quarantine. The recipient institution, while gaining access to the material, also would accept responsibility for performing tests to detect pathogenic agents and for developing pathogen-free stocks for distribution. However, facilities located within the primary agricultural region for a crop may not be appropriate for quarantine.

Greater use of the option to release quarantined germplasm to qualified scientists, with the provision that it remain under quarantine, should be considered. This would apply only to materials being introduced for scientific purposes and not to commercial-sized lots.

*More use should be made of overseas, third-party, or isolated offshore quarantine facilities.*

Isolated or third-party quarantine facilities outside the United States that are developed with other nations or international centers provide good alternatives for supplementing present facilities or to access environments more suitable to the growth of the plants being introduced. The testing of material in a non-U.S. location would be acceptable if it meets scientific and quarantine criteria. Stations in Europe, for example, could serve well for the quarantine of certain kinds of material. The location should be one where the tested crop is not grown and endemic hosts for its pathogens are not present. Environmental, biological, and political aspects must be considered in developing cooperative arrangements. Benefits to both the United States and the cooperators must be clearly established and set forth.

The NPGS provides third-party quarantine services for other institutions. The Centro Internacional de Agricultura Tropical (CIAT, International Center of Tropical Agriculture) in Colombia, for example, uses the Western Regional Plant Introduction Station to quarantine bean germplasm from Africa. The CIAT is an international research center in the CGIAR system.

*The National Plant Germplasm System should work with the Animal and Plant Health Inspection Service, which is responsible for quarantine programs, to support research on technologies for rapid and reliable detection and elimination of pests and pathogens.*

Selected plant species are placed in quarantine when they enter the United States to prevent the introduction of pests or diseases that could harm U.S. agriculture. This requirement is often seen as an impediment to germplasm acquisition. Most imported plants pass through quarantine following little more than visual inspection, but in some cases (e.g., *Prunus* species) quarantine can entail years of isolation and testing before the germplasm is released. The NPGS has developed priorities for the materials and agents of greatest concern to guide research on quarantine, but has limited funds. Molecular technologies now emerging hold the potential to greatly reduce quarantine periods, particularly for detecting viruses or mycoplasmas. Methods should be further developed and adopted to screen for intracellular organisms.

The protocols developed for molecular identification should be rapid, definitive, and relatively simple to use to enable the screening of greater numbers of entries. While the technologies for developing such methods are available, their application to disease agents of particular significance has been slow.

The development of pathogen-specific protocols for routine use in quarantine is needed. Chemotherapy, thermotherapy, and meristematic tissue culture are examples of such procedures. The required research could be accomplished by specialists in cooperation with scientists at USDA facilities.

## National Collections

*Collections must be managed as national, not regional, resources.*

If germplasm activities are to have a national focus and international recognition, collections should be fully integrated into a national system. The management of that system should be coordinated at national rather than regional or area levels. For example, all of the NPGS sites could be designated as national plant genetic resources centers, and individual collections could be designated as national plant genetic resources collections.

These or similar designations will underscore the national focus for germplasm activities. While many of the national system's sites arose as regional efforts, they must now have stronger national and international significance. The national germplasm collections could be managed uniformly, with budget and administrative oversight exercised by one central office. The designation "national collection" would carry with it the requirements for certification and adherence to basic management standards, and the assurance of basic support to ensure safekeeping.

## Crop Curators

*Curators with specific knowledge should be appointed for each major crop or crop group, and they should be given management responsibilities.*

Knowledgeable curators must oversee the acquisition and management of all of the major or essential NPGS collections, and promote their use. The lack of many such curators requires some site managers to oversee a number of different crop species. Curators must have specific knowledge about their crops, collections, maintenance sites, needs, and enhancement plans. With advice from the crop advisory committee, a curator should work with the leader of the NPGS to develop and implement plans for exploration, management, documentation, regeneration, evaluation, and enhancement. By fostering the breeding of potentially useful genes into appropriate genetic lines, curators can greatly enhance the use of NPGS collections.

## Base Collections

*The base collections at the National Seed Storage Laboratory should reflect all of the seed collections in the national system.*

The NSSL is intended to provide base storage for all materials held in the national system's seed collections, but it cannot because of space limitations. Expansion of the laboratory is therefore imperative (National Research Council, 1988). However, as collections enlarge, the size of the NSSL's task will increase. The development of improved protocols for storing seed, monitoring its viability, and regenerating small or declining samples is imperative.

*The National Plant Germplasm System must devote more of its resources to regenerating seed accessions.*

Regeneration is costly in time and resources, and introduces the risk of genetic shifts, particularly where an accession is genetically heterogeneous. It is, however, necessary if seed viability has declined, or if distribution or testing has depleted the accession.

Regeneration must take account of the breeding structure and population genetics of the sample. Errors, such as the lack of cages or inadequate separation distances to prevent cross-pollination of open-pollinated accessions in the field, can result in irretrievable genetic damage. For some accessions, unique, non-native-pollinating insects may need to be maintained with the accession.

In the past, resources to increase or regenerate such samples have been inadequate. For about 60,000 NSSL accessions, mostly named cultivars without PI numbers, no site has been identified where samples could be sent for regeneration. The NPGS needs to develop long-range plans for regenerating accessions with few seeds and those of low or declining viability. Where responsibility for providing fresh seed cannot be assigned to an NPGS site or collection, funds should be available to the curator to secure regeneration on a contract basis, with appropriate supervision and safeguards. In a survey of tests conducted at the NSSL over the past 10 years, the committee found that 71 percent of the laboratory's collection was above 65 percent germinability (see Table 2-10). However, 45 percent of the samples contained less than 550 seeds. Regeneration of these samples is urgently needed.

Regeneration activities will increase over time, and as collections enlarge and more NPGS material is stored as backup at the NSSL. Technologies to decrease viability loss such as cryopreservation (storage in or suspended over liquid nitrogen at temperatures between −150°C and −196°C) should be pursued. Methods of identification are needed to ensure that seed received after regeneration is genetically the same as that sent out.

Seeds are kept in cryogenic storage at the National Seed Storage Laboratory. Credit: U.S. Department of Agriculture, Agricultural Research Service.

## Special Collections

*A plan should be developed for monitoring, supporting, and conserving important special collections.*

There are many collections outside the federal system that vary in their taxonomic or geographic emphases, completeness, and scope. Many were assembled by university, state, or ARS scientists. Others were developed outside the regional station system, or were begun prior to its formation. Some accessions are genetically unique, and others are similar to those already held by the national system. The extent of duplication between these collections is largely unknown.

It is difficult to ascertain who is responsible for these collections. The intermingling of commodity, germplasm, financing, and other program responsibilities at ARS has made it difficult to manage these collections as parts of the NPGS. Some collections are the responsibility of an individual with narrow and specific commodity interests. The leader of the NPGS and the appropriate crop advisory committee should have access to information about the status of these collections.

A management plan for these special collections should cover four points. First, many collections still need to be identified. At least three are known to have been partially lost due to lack of a timely transfer of materials or mishandling after transfer (i.e., the Mangelsdorf maize collection, the Stephens cotton collection, and the Whitaker *Cucurbita* species collection). Second, such collections need back-up storage at the National Seed Storage Laboratory. Third, unique collections should be duplicated and made a part of the NPGS. Fourth, the maintenance status of important collections outside the national system and held by individuals should be monitored to provide insurance against their future loss.

The collection of tomato species at the Charles M. Rick Tomato Genetics Resource Center is an outstanding example of an important special collection (Genetic Resources Conservation Program, 1988). This collection is maintained at the University of California at Davis, with partial support from ARS through special funding arrangements. Most (97 percent) of the available accessions at the Rick Center are duplicated at the NSSL.

### Genetic Stocks

*The National Plant Germplasm System should provide secure, long-term storage at the National Seed Storage Laboratory for genetic stocks and should assist in the support of collections that are considered important for agriculture or basic research.*

Genetic stocks typically possess one or more genetic anomalies (e.g., multiple or missing chromosomes, unique genetic markers) that make them of interest to researchers. The management of these stocks involves a continuing search and examination for particular genetic traits or anomalies. Maintaining genetic stocks can be complicated by mutant genes or chromosome aberrations that reduce the viability of their seed. Mutant stocks often can only survive under very specific conditions and frequently require highly specialized procedures to multiply, identify, and maintain them. Genetic stocks are therefore not generally held in active germplasm collections, such as those of the regional stations.

There is an acute awareness of the importance of genetic stock centers as essential underpinnings of basic and applied research and education on plants both in the United States and throughout the world (McGuire and Qualset, 1990). These collections have been and are important for plant breeding, biosystematics, genetics, development, physiology, biochemistry, and molecular genetics. Advances in knowledge that came through the use of materials in these collections include the regulation of gene function, processes of genetic mutation, fine structure of genetic material, behavior and mechanics of chromosomes, starch and storage protein biosynthesis, and the existence and properties of transposable elements (migrating genetic materials). Stocks with multiple markers are frequently requested for linkage analysis studies.

Over the years, genetic stock collections have been financed by a combination of National Science Foundation, USDA, state, or other funds on an ad hoc basis. With the current emphasis on molecular biology and plant genome mapping, such collections have become even more important, yet ARS has no program for these collections, and the NPGS is only beginning to develop policies for handling them when they become endangered. Efforts are being made to provide secure conservation for selected collections and important stocks. However, many individuals who maintain important barley, corn, tomato, and wheat genetic stocks are at or past retirement age; few replacements are in sight. Furthermore, there appears to be no plan to encourage or engage other suitably qualified scientists to replace them.

Because of the specialized nature of mutant stocks, the difficulty of deciding what to maintain or expand, and the expense that maintenance could entail, a decision must be made about the extent to which the NPGS should be involved in their maintenance and elaboration. The Crop Science Society of America has adopted a recommendation that genetic stocks of significance be registered and samples of those stocks be deposited in the NSSL (White et al., 1988). This policy was encouraged by the NPGS. The NSSL is able to store most genetic stocks, but their management, characterization, regeneration, and distribution should remain the responsibility of the knowledgeable scientists actively studying and using them.

The NPGS should have the capacity to provide a limited amount of supplemental funding for selected critical collections, including mutant genetic stocks important to agriculture and biology. Funding, on a limited basis, should be available to provide a safety net against loss of important collections. First priority should be given to the rescue of orphaned and endangered mutant genetic stock collections.

### Establishing Core Subsets

*The management and use of large collections, such as those for wheat, corn, and soybeans, could be aided by the identification of core subsets, but this method must be applied cautiously.*

As collections of plant germplasm around the world have grown over the past 25 years, so has the magnitude of the task of managing them. Concerns have been raised that many major crop collections, such as wheat, barley, and rice, have grown so large and diffuse that they inhibit, rather than promote, effective management and use (Brown, 1989a,b; Holden, 1984). Increased emphasis on collecting wild and weedy relatives and expanding genetic coverage of present collections has created the potential for collections to grow even larger. While it can be argued that collections of large numbers of accessions are indeed valuable resources (Chang, 1989), their management can be difficult with only limited resources available.

Removal of duplicate accessions in a collection can reduce maintenance costs (Chang et al., 1989), but it is unlikely to have a significant impact. While the identification of duplicates is conceptually simple, the accuracy of technologies (e.g., restriction fragment length polymorphism analysis and protein electrophoresis) in identifying duplicate accessions in larger collections is still a matter for debate. This is especially true for crops to which such technologies have not previously been applied. Further, applying the newer molecular technologies to a large collection can be time consuming and expensive.

The problem is not so much to remove redundant materials as it is to enable efficient management and use of the wide range of diverse materials in collections. This has led to proposals for developing core collections worldwide. To underscore the fact that these are not separate collections, but a set of designated accessions within an existing collection, they are here referred to in this report as core subsets.

The core concept entails identifying a range of accessions within a collection, the total of which would include, with an acceptable level of probability and with minimum redundancy, most or much of the range of genetic diversity in the crop species and its relatives (typically no more than 10 percent of the whole collection) (Brown, 1989a,b; Frankel and Brown, 1984). The goal of this subdivision is to facilitate use, and in particular, to provide efficient access to the probable range of variation in the whole collection in which a scientist may have interest (Brown, 1989a,b). It is envisioned that a breeder or researcher seeking particular traits could first examine a well-defined core and, if necessary or desirable, use the results of that search for a precise examination of the collection on the basis of passport or characterization data.

The core subset should contain the breadth of genetic diversity available for a crop. However, it is probable that if 10 percent of the total accessions are assigned to a core subset, no more than 70 percent of the alleles in the whole collection will be included (Brown, 1989a,b; Strauss et al., 1988). The entire collection, therefore, must continue to be maintained. It serves two crucial functions: first, as a source of additional diversity that may be similar to the core, but due to different genes or allelic combinations could differ appreciably from core germplasm; and second, as a source of genetic variation for traits that are not represented in the core subset. While use of a core concept for managing a collection might shift the priority for evaluating certain accessions, it does not provide information for eliminating, bulking, or combining accessions.

The core subset has advantages for curators. It would be more widely and readily distributed than the whole collection. It is also expected that the core would receive priority in evaluation and characterization. Thus, curators could make more efficient use of limited budgets through directed efforts, promote the distribution of their materials and information, and facilitate use of the collection.

It should be emphasized that core subsets are management and record-keeping tools. They guide activities and the development of priorities for evaluation, characterization, and distribution. They do not require the physical construction of separate facilities or storage areas. They do, however, require availability of a basic amount of passport information for each accession in a collection. This basic information enables the individual accessions in the core subset to be linked to potentially wider diversity in the entire collection.

Implementation of the core subsets concept in NPGS collections is presently constrained by practical and scientific difficulties. The concept requires that much basic information be available on accessions, but many accessions lack such information. Methods for selecting a core subset are still a matter of scientific debate (Brown, 1989a,b; Chapman,

1987; Strauss et al., 1988). Current proposals are based on sampling theory and population genetic theory, but they are as yet untested in large collections.

Core subsets for some of the larger NPGS collections such as wheat, barley, and beans, could aid curators in management and exchange, providing sufficient documentation was accessible. Although unlikely to reduce the size of present collections, they might reduce the frequency of acquiring duplicate or near-duplicate samples, and increase the ease of access to the collection by simplifying the screening process for particular traits. The danger, however, is that this approach could be seen as a mechanism for reducing collection size, combining materials, or simply neglecting germplasm that is not part of the core. For collections that already are small or that represent limited geographic distribution, a core subset could be too small to represent the collection's diversity, thus being an inappropriate management strategy.

## Cooperation with Other Collections

*The National Plant Germplasm System should seek opportunities for cooperation with other efforts to preserve plant germplasm.*

Collections of heirloom and specialty crops may hold germplasm not found in NPGS collections (Office of Technology Assessment, 1985). They are resources for which little USDA maintenance effort is required, and are sources of old varieties no longer available and not now part of NPGS collections. Private groups that maintain heirloom and specialty crops should have access to the expertise of NPGS scientists and backup facilities for their collections. For example, through a memorandum of understanding with the Center for Plant Conservation (CPC), the national system provides base collection storage of the seeds of rare plants held by CPC's cooperating institutions. Communication with individuals from private interest groups could be improved by their membership on relevant crop advisory committees.

Botanical gardens and arboreta hold large numbers of species not found in NPGS collections. These are often very small, or single plant, samples from natural populations and are not genetically diverse. It would be desirable for the NPGS to cooperate to a greater extent with botanical gardens in evaluating ornamental species, as is done at the U.S. National Arboretum. With the help of botanical gardens in the CPC consortium, the NPGS could evaluate plant materials in a wide range of regions and habitats. Efforts along these lines have been made at the regional station in Ames, Iowa.

The national system also could provide support through special grants for activities such as the *Garden Seed Inventory* of the Seed Savers Exchange (1989a). This is a potentially valuable resource for monitoring the genetic diversity of commercially available vegetable seed. A minimal level of support from the NPGS could assure its continuance.

Cooperation could reduce the tensions that occur when private groups must compete with the national system for limited quarantine resources. Space for importing materials that must be quarantined for extended periods is limited and access is provided to the NPGS on a priority basis. This could lead to APHIS refusing requests for imports by other interests (Yazzie, 1989). Cooperation with groups such as the Seed Savers Exchange or CPC could lead to more equitable agreements and help to reduce unauthorized importation of plants.

Companies with access to subtropical sites with warm winters and short day-lengths are particularly useful for regenerating certain species. Their nurseries, located in the southern United States and abroad, can readily facilitate large-scale increases of seed. The Latin American Maize Project—a cooperative venture with private financing that involves U.S. universities, several Latin American countries, and the USDA—is evaluating seed stocks from a wide array of Latin American maize landraces (White and Briggs, 1989).

Demands for exotic, wild, and weedy species related to crops are expected to increase as molecular technologies for transferring alien genes to crop plants become readily usable. The search for new characteristics requires the availability of germplasm beyond that contained in current cultivars and breeding lines. The biotechnology industry should be encouraged to support the activities of the national system. The amount of university and industry biotechnology research projects that complement USDA's germplasm activities should expand and enhance the use of a broader array of genes from a widening base of germplasm.

### In Situ Conservation

*In situ conservation methods should be used to complement the primarily ex situ activities of the National Plant Germplasm System.*

These methods include maintenance of wild plant genetic resources where they occur naturally or maintenance of domesticated materials where they were originally selected. For many species native to the United States, such as blueberries, cranberries, or *Rubus* species, in situ conservation may be accomplished through the designation of existing parks, wildlife refuges, or other protected areas as in situ reserves. The status of species or populations conserved in these areas would have

Wild barley, wild oat, and wild wheat are grow at an in situ research site at Ammiad in the eastern Galilee region of Israel. Credit: Calvin Sperling.

to be monitored and the information could be maintained in the Germplasm Resources Information Network (GRIN). Cooperation with groups, such as Native Seeds/SEARCH (Southwestern Endangered Aridland Resource Clearing House), that are already involved with in situ monitoring of selected species might also be arranged.

Efforts should be made to promote in situ conservation for plant genetic resources outside the United States. For example, financial aid through programs of the U.S. Agency for International Development (USAID) or technical assistance coordinated by the NPGS could be provided to enable developing nations to conduct surveys, identify regions, or establish appropriate in situ reserves or activities. Where aid is provided through another agency or department, the advice and technical expertise of the NPGS should be sought. Funds, for example, could be provided by USAID for the in situ conservation of *Hordeum spontaneum*, a barley wild relative, and *Triticum dicoccoides*, a wheat wild relative, in Israel, and technical support could come from the NPGS.

## FACILITIES AND PERSONNEL

The national system's germplasm is housed in collections at many sites in the United States. The use of multiple sites are necessary because of the diverse environments needed to maintain all of the accessions, but it makes overall management of the system difficult. Many accessions require specialized environments not available at the given maintenance site or within the system.

### National Seed Storage Laboratory

*The National Seed Storage Laboratory must be expanded.*

Several reports have stressed the importance of upgrading the physical facility at the NSSL (Council for Agricultural Science and Technology, 1984; General Accounting Office, 1981a,b; National Research Council, 1988; Office of Technology Assessment, 1987; U.S. Department of Agriculture, 1981). Expanded facilities should provide significantly more space for the storage of accessions, for the separate storage of quarantined germplasm, and for collections held as backup for other national and international institutions. The committee endorses these earlier reports.

The securing of an appropriation for this expansion must continue to be a high priority for the USDA. A federal restriction prevents construction until appropriations sufficient for completion have been received. The combination of an aging facility and insufficient space make it increasingly difficult for the NSSL to serve the purposes for which it was built. Space constraints have forced the NSSL to develop plans for temporarily storing selected materials at alternate sites, which lack long-term storage capacities (S. A. Eberhart and H. L. Shands, U.S. Department of Agriculture, personal communication, August 22, 1990).

Coupled with expansion of the NSSL must be an increased regeneration of samples at regional stations or other locations and additional funds to support this activity. Where an NPGS maintenance site does not exist, funds must be available to contract for and inspect grow-outs.

### External Review

*Facilities and programs of the National Plant Germplasm System should undergo periodic external review.*

National germplasm sites and crop-specific collections should be reviewed regularly to ensure that their programs, resources, and staff capabilities meet the needs of the germplasm for which they are conservators. Reviewers should have technical or management experience in plant germplasm or its use, and should be drawn from outside the national system. The appropriateness of a site's location should also be considered. While it is not advisable for collections to be moved frequently, the location of collections should be based on scientific considerations and opportunities for cooperation with universities or experiment stations. The selection of reviewers, the development of guidelines, and the evaluation of results and their distribution should be the responsibility of the leader of the NPGS, with direction and advice from the Plant Germplasm Operations Committee, a research advisory committee, or an ad hoc committee assembled to oversee the external review of a site.

## Location

*The locations of national system sites should be reviewed and where needed alternate locations should be identified.*

Existing clonal repositories were established on the basis of regional interest, growing conditions, lack of pathogens, and local expertise. A wide range of state, federal, and private sector experts provided advice to the site selection process. The risk of maintaining plants that may harbor pathogens not yet found in the region was considered. Further, some collections have proved difficult to maintain in the presence of specific pathogens that have been found to exist in a region. Eastern filbert blight *(Anisogramma anomala)*, for example, presently occurs within 50 miles of the Corvallis, Oregon, repository and thus may hamper maintenance of this germplasm at that site.

Close collaboration with local institutions is essential and must be recognized. The physical proximity of NPGS facilities to university campuses and state agricultural experiment stations is an asset. So, too, is the sharing of staff and program activities. However, there should be no ambiguity about the primacy of national policy and authority over managing collections, or using resources and personnel provided for that purpose. Where support from an experiment station has declined or maintenance has been inadequate, the possibility of moving a collection to a more suitable location should be considered.

## New Facilities

*Sites should be established for the growth and maintenance of germplasm that requires short day-lengths or arid environments.*

The management of accessions with physiological and environmental constraints related to their maintenance can be difficult. These are plants that cannot be reproduced because day-length or temperature requirements cannot be met in the United States. Some accessions requiring warm, short days are amenable to winter culture at facilities in Puerto Rico or St. Croix, but for other reasons cannot be grown at these sites. Some short-day beans, from the Regional Plant Introduction Station in Pullman, Washington, for example, will flower at these Caribbean stations, but native insects increase cross-pollination among accessions and the difficulty of maintaining genetic integrity. High-elevation materials often require cooler temperatures (especially cooler nights) than available in tropical or subtropical island habitats. Controlled greenhouse facilities have been used to meet some of these needs, but they are very costly. Controlled facilities do, however, provide security from the often unpredictable hazards (e.g., pests, disease, weather extremes, unanticipated pollinators) that can make field increase a risky endeavor. The need for an adequate population size to preserve genetic diversity can constrain the use of controlled environment facilities for regular regeneration.

Facilities at which short-day responsive plants can be grown and multiplied in the field should be found. They are required for a wide range of annual and perennial plants that must be regenerated, characterized, and evaluated. Field maintenance and growth would greatly increase the capacity for these activities over that of greenhouses. The site should be frost-free throughout the year, with low relative humidity, freedom from violent storms, and at a sufficient altitude to avoid extremes of heat. It must be accessible to irrigation and sufficiently isolated from general agriculture to provide a relatively disease-free environment. A site outside the United States that meets these requirements could be established by cooperation among USDA, USAID, and IBPGR. Adequate monitoring and precautions must be taken at such sites to prevent contamination of regenerated accessions with pests or pathogens that could be inadvertently spread when seed is distributed.

The United States also needs a facility located in an arid region, such as the drier desert southwest where pests and disease are typically less abundant. Cooperative agreements already provide for regenerating cotton and some small grains in this region. An NPGS site would be particularly valuable for maintaining selected accessions of sorghum, bean, grass, small grains, and jojoba.

### Staffing

*The federal and state partnership for managing plant germplasm should be reappraised and reinforced.*

Because it is sometimes difficult for ARS employees to supervise state workers directly, the ARS has tried to bring the staffing of some sites under the control of its personnel system. This has led to "federalizing" sites, such as the National Clonal Germplasm Repository at Corvallis, Oregon. As a result, the system has been perceived as drifting away from the original concept of shared responsibility and support.

Continuing the association of federal scientists with state university colleagues is important because it permits access to a broad array of expertise. Germplasm sites require staff trained in agronomy, horticulture, plant pathology, genetics, entomology, and plant systematics. Achieving this disciplinary mix has not been possible at all sites, and the needs of many germplasm centers have only been partially filled. More recently, staffing has been increasingly controlled by ARS. This shift has come about in part because of declines in CSRS or host university support relative to current costs. Efforts are needed to renew and strengthen the cooperative environment that has made available much needed expertise.

Individual interests of scientists at NPGS sites with specialized research responsibilities occasionally have conflicted with germplasm maintenance needs. In general, the committee has been favorably impressed with the focus, energy, expertise, and enthusiasm of many NPGS staff. However, many valuable individuals could become discouraged and leave unless actions to provide suitable opportunities for advancement and career development, while promoting the management of NPGS germplasm, are taken.

## Training

*A clearly defined policy for training scientists and technicians in the management of germplasm should be developed and implemented.*

As the activities of the NPGS grow, so will its need for new professional and support staff. Further, as a partner in managing the world's genetic resources, the United States has an obligation to provide training opportunities for international scientists. By training international scientists, the NPGS can open opportunities for closer cooperation in managing and exchanging genetic resources. The best mechanism is site-based training.

While the NPGS is in an ideal position to train new professional and support staff, little coordination exists to guide and direct training. Opportunities for providing on-site training are not utilized at some sites. Central administration of training would enable more efficient use of limited staff time and facilities. Both long-term training and study opportunities for scientists and short-term workshops or in-service training are needed.

Training takes a significant amount of time and resources, both of which are limited. However, the committee found that for many NPGS sites the existing level of training is far below the present capacity to provide it. Many sites would not need larger staffs or increased budgets to support training. It should also be possible as well as advantageous to coordinate training with the academic programs of institutions where NPGS facilities reside.

In-service training is a necessary and an integral part of the NPGS. When sites are developed or enlarged, their potential to provide training should be considered to ensure that suitable space and facilities exist. The committee's report on the National Seed Storage Laboratory included recommendations for training facilities (National Research Council, 1988). Improving these facilities should not add substantially to construction costs when done as part of major expansion or remodeling efforts.

## Classification and Promotion

*The system of classifying and promoting scientists should be reexamined and adjusted to not only reward, but attract and retain career scientists needed for germplasm work.*

ARS research scientists within the NPGS are judged primarily on the basis of published, peer-reviewed research. However, the majority of their work is the evaluation, regeneration, and record-keeping necessary for good germplasm management which may not lead to published papers. A system of classification and promotion that attracts, rewards, and retains career germplasm scientists is needed. It should take account of the considerable time required of NPGS scientists to perform tests, examinations, and other work not related to publication. It must not, however, produce barriers that discourage scientific study and publication in areas related to germplasm and its management.

Germplasm management is now recognized as essential to plant and animal improvement, to conservation programs, and to successful food production. However, there is no scientific classification within ARS commensurate with the significance of this activity. An appropriate system of advancement should be developed that enables adequate recognition of both the service and goal-directed research activities unique to germplasm work.

If the NPGS remains within the ARS, changes in personnel classifications and criteria for evaluation will be needed. One option would be to restructure the evaluation system so that reviews of germplasm scientists and site managers are based on accomplishments in managing collections. Records or evidence of germplasm-related accomplishments, such as evaluation or regeneration, should be considered in addition to published papers. Alternatively, the ARS could create a separate category for germplasm scientists that allows for open-ended promotion, but with evaluation guidelines more appropriate to the work of germplasm maintenance and research. The potential to structure a staff appointment to devote a percentage of time to basic research (e.g., 80 percent for service activities and 20 percent for research) should not be overlooked.

## THE MISSION OF THE NATIONAL SYSTEM

*The National Plant Germplasm System should develop clear, concise goals and policies that encompass the conservation of plant genetic resources that reflect the world's biological diversity and crop resources of immediate use to scientists and breeders.*

Efforts are needed to expand some collections to make them more representative of the available diversity. Assessments of collection completeness must give due regard to the inclusion of close wild relatives and non-crop-related species, which may possess useful genes. National policy should include endangered species of native and exotic taxa and should not be limited to crop genetic resources.

Plans for collecting should include consideration of the range of ecogeographical areas where accessions originate, how broad based or narrow the collection is in terms of known or suspected genetic traits, and what genes might be obtained through various transect or other sampling procedures when rare alleles are sought. These factors must be weighed against cost, accuracy, need, and other criteria for obtaining suitable materials.

Specifically, definitions and plans are needed for

• Developing a long-term policy, periodically reviewed and revised, that states what genetic resources will be acquired and how to cooperate with foreign germplasm banks and with U.S. collections that are not formally part of the NPGS.

• Assessing NPGS collections and developing priorities to ensure they sample adequately the genetic diversity for the species.

• Replenishing seed stocks, with the help of international collaboration where appropriate.

• Characterizing and evaluating the germplasm held in collections. This information will facilitate wider use of germplasm and make possible more efficient management.

• Accelerating the adoption of modern technologies for the maintenance and characterization of germplasm.

- Promoting the use of conserved germplasm through enhancement efforts that incorporate important new genetic traits into appropriate genetic backgrounds.
- Clarifying the position of the United States on questions of germplasm ownership, unrestricted exchange, and cooperation that are emerging in international forums. The committee endorses an earlier recommendation (Office of Technology Assessment, 1987) that the United States should not embargo exchanges of germplasm for political, economic, or other reasons.
- Developing cooperative links between the NPGS and other national and international agencies, institutions, or groups conserving global biological diversity.

*The United States must address the problem of the global loss of biological diversity. This can be done in significant part through conserving the genetic diversity of crop species.*

The stated mission of the national system is "to acquire, maintain, evaluate, and make readily accessible to crop breeders and other plant scientists as wide as possible a range of genetic diversity in the form of seed and clonal germplasm of our crops and potential new crops" (U.S. Department of Agriculture, 1981). The NPGS collections are not storehouses for all of the known crop cultivars. Other groups hold and distribute heirloom varieties no longer commercially available.

While the focus of the NPGS on breeders and researchers as primary users has been satisfactory to date, the need to conserve the world's biological diversity has become an important issue. Aggressive participation and leadership by the United States in international efforts toward conserving and managing biological diversity are necessary and urgent. In the future, national programs could include broader biological conservation and research interests, such as conserving threatened or endangered wild species unrelated to crops but of potential economic or unique biological interest (Office of Technology Assessment, 1987) or in situ monitoring of the status of crop progenitor species and landraces threatened by habitat decline.

## International Policies and Cooperation

*The National Plant Germplasm System must take a more active role in developing U.S. policies that guide relations with the Food and Agriculture Organization, international agricultural research centers, and other international agencies and national institutions.*

Germplasm management is no longer a strictly national concern. By accepting responsibility for international base collections, the U.S. underscores its international role in maintaining these resources, for itself and for other nations. To fulfill these obligations requires a coherent, science-based policy. To achieve it will require the NPGS and its leadership to have a much greater role in developing U.S. policy related to germplasm and in representing the United States before international germplasm bodies, such as international agricultural research centers and the FAO.

While individuals in the NPGS may cooperate and interact with their counterparts in other nations, there is no U.S. international policy on plant genetic resources. Much of this is due to the domestic focus of the Agricultural Research Service. Many individuals within the ARS have developed international contacts, but such cooperation is not part of the service's policies or goals and there is limited support for it. The USDA generally defers to the U.S. Department of State on international matters. When there is a lack of appreciation or understanding of the scientific and agricultural issues involved, the outcome may not promote germplasm interests.

Cooperation with other nations in managing genetic resources should be more widely pursued. Greater exchange of germplasm and data between the NPGS and similar institutions in other nations (e.g., N. I. Vavilov Institute of Plant Industry in the USSR, National Institute of Agrobiological Resources in Japan, Institute of Crop Germplasm Resources in the People's Republic of China, and the National Bureau of Plant Genetic Resources in India) will lead to other mutually beneficial cooperative efforts. While some efforts at international cooperation exist, there is no clearly established mechanism or policy for fostering them.

There should be an unambiguous mechanism for establishing U.S. positions with regard to germplasm. In the past, it has been unclear what office is responsible for making recommendations to international forums or for defining U.S. actions regarding the management of genetic resources. Opinions expressed by scientists, administrators, and advisers seemingly disappeared in the interagency bureaucracy. The NPGRB, as an adviser on germplasm issues, could take the lead in discussing these issues. However, there must be a mechanism for acting on its recommendations.

*The United States should become a member of the Commission on Plant Genetic Resources of the Food and Agriculture Organization of the United Nations.*

The United States does not participate in the FAO Commission on Plant Genetic Resources as a member, nor is it a signatory to the FAO

Undertaking on Plant Genetic Resources. The commission is an important forum for discussing international issues related to cooperation, exchange, and ownership of genetic resources, particularly with developing nations. Until recently, the NPGS was not officially represented in the observer delegation sent by the United States to commission meetings.

The NPGRB has only recently begun to discuss the possibility of U.S. participation in the FAO Commission on Plant Genetic Resources. It passed a resolution on November 14, 1989, recommending that the United States become a member of the commission. In August 1990, the Bureau of International Organization Affairs of the U.S. Department of State notified USDA that it concurs. No actions have been taken or recommended by the NPGRB in regard to adhering, completely or with reservations, to the International Undertaking on Plant Genetic Resources or supporting an international fund established by FAO to support the genetic resources activities of commission members.

Membership in the FAO commission does not imply agreement with all of its policies or positions. Because it is possible to join the commission without adhering to all or part of the undertaking, the United States could underscore its concern for genetic resources and gain a voice in this important international forum. Membership would enable the United States to help in shaping the agenda and activities of the commission. [The United States joined in September 1990.]

*The National Plant Germplasm System should cooperate with other nations to conserve, collect, maintain, and regenerate germplasm.*

U.S. germplasm activities have been largely guided by an unofficial policy of national self-sufficiency that calls for comprehensive collections to reduce dependence on other nations or collections. This policy frequently does not recognize the increasingly international nature of germplasm acquisition, management, and conservation, and the necessity to foster global cooperation. Other collections outside the United States not only hold important germplasm resources but can provide vital support to the NPGS. For example, cooperation on seed regeneration could allow a division of labor and costs among nations or institutions without sacrificing national self-interests.

Many nations, especially those with rich genetic diversity, are reluctant to allow collection and exchange by foreigners. The United States must seek to preserve open and unrestricted exchanges of germplasm. The United States has international base collection responsibilities for maize, but there are no U.S. experimental facilities suitable for regenerating accessions of high-elevation, short day, Andean maize landraces. Agreements with other nations are needed for regenerating germplasm materials for which suitable environmental conditions do not exist in this country.

*The United States should expand its support of cooperative activities with the international agricultural research centers affiliated with the Consultative Group on International Agricultural Research.*

In 1989, the United States, through USAID, provided about $42 million of the annual budget ($272 million) for core operations of the CGIAR centers and contributed additional funding for special projects. Only a small portion of these funds support germplasm activities. Individuals within the national system work cooperatively at times with scientists in CGIAR institutes (e.g., the NSSL holds partial duplicate collections of rice, maize, wheat, and beans), but there are no formal agreements or coordination of germplasm activities between the research centers and the national system.

In particular, greater collaboration between the NPGS and the IBPGR should be encouraged. The sharing and exchange of computerized inventories and data on germplasm could enhance the sharing of responsibilities between international institutions and the national system for managing many large collections. Back-up storage at the National Seed Storage Laboratory of the rice collection from the International Rice Research Institute and the cooperation of the NPGS in designating many of its collections as international base collections are examples of cooperation.

There is little formal sharing of computerized passport, characterization, or evaluation data between the NPGS and the CGIAR research centers, and no real attempt to standardize data records between them. Many individual scientists and research facilities have developed working relationships with counterparts at the research centers. For example, the regional station in Pullman, Washington, cooperates with the CIAT in Columbia on beans and with the International Crops Research Institute for the Semi-Arid Tropics in India on chickpeas. The North-Central regional station cooperates with the Centro Internacional de Mejoramiento de Maíz y Trigo (International Maize and Wheat Improvement Center) in Mexico on maize. There is, however, little formal interaction between the NPGS and international organizations.

The NPGS and the CGIAR's research centers develop descriptors for characterization and evaluation of germplasm independently. The United States should seek to use descriptors and other data that are compatible with the centers to improve exchange of information on collections.

*The United States should work with neighboring countries to establish a North American cooperative program in genetic resources.*

Both Canada and Mexico have national plant germplasm systems. The NPGS has cordial relations with the genetic resources staffs of both countries. Canada, Mexico, and the United States could benefit from closer linkages and access to materials held by each. Responsibility for specific collections could be shared among them. Other advantages include cost savings for the participants; facilitation of regeneration, germination testing, and quarantine; expansion of the total range of materials accessible to each nation; and increased security, provided by backup for selected collections. As part of this cooperation, representatives of the Canadian and Mexican germplasm programs could hold ex officio membership in the NPGRB.

### INFORMATION MANAGEMENT

The NPGS has a total of about 380,000 accessions. However, many computer records are incomplete and lack even the most basic descriptive information. Increased funding has made possible some updating of information on priority accessions, but this is a continuing process and must be sustained. The cooperation of all users of NPGS germplasm is essential to gather information about accessions. Recipients of germplasm should share their evaluation and other data with the NPGS.

The system's information, which covers collection inventories and germination records to evaluation data and exchange requests, is managed by the Database Management Unit (DBMU) using the Germplasm Resources Information Network. As noted earlier, the database was set up to serve as a central information repository, begin the standardization of crop descriptors and evaluation information, and help curators manage collections. These are three very different functions. The first entails obtaining and storing detailed information about accessions and requires sophisticated data retrieval capabilities. The second is a data classification activity for uniform and efficient data handling. The third is inventory control to help collection managers. The problem is that in developing a database system, modifications that help one function may hinder the others. The committee questioned whether it is appropriate to pursue a single, all-encompassing system to accomplish these differing tasks. For example, it is possible to develop standardized descriptors, but much of the available entry data may not be expressed in a form compatible with them. In fact, GRIN record formats for individual accessions contain lists of crop descriptors for which little, if any, data have been entered.

**Completion of the Database**

*The Germplasm Resources Information Network must better reflect the collections of the National Plant Germplasm System.*

An accurate directory, or central database, of all of the holdings of the national system is essential. The GRIN was intended to be a central data facility. While the structure and operation of the database management system are in place, the NPGS has experienced difficulties and delays in locating, correcting, and loading data that accurately represent its holdings. Access to a directory will make it possible to determine what materials are in the collections and where they are located. A directory would be a valuable tool for establishing priorities and focusing activities.

The GRIN includes inventories of the PI numbers and of several site collections, such as holdings of the NSSL and the clonal germplasm repository at Miami, Florida. The site inventories include items without PI numbers and materials that are not considered part of the national collections. Thus, no single set of inventory numbers can be used to assess the accessions of the NPGS. Further, the GRIN database still does not contain all inventories of all sites. Accession records for much of the clonal germplasm, for example, remain to be entered. Thus it is not yet possible to determine the total inventory of the NPGS.

Individual sites generally have inventories of all of the germplasm that they hold. The NSSL, for example, has conducted an inventory of its holdings to confirm their active status, reduce duplication, and eliminate redundancies. Holdings listed in other collections within the system can range from diverse accessions for a particular species to germplasm represented by only a single accession. For example, the list of germplasm holdings provided to the committee by the National Clonal Germplasm Repository at Mayaguez, Puerto Rico, contains several listings of single accessions representing a species. The national germplasm collections should contain those plant materials held for the purpose of providing a broad, accessible germplasm base. Individual plants may be the only germplasm source and may point to the need for additional material if the species has economic or other value.

**Passport and Descriptor Data**

*A high priority should be placed on completing the listings on the Germplasm Resources Information Network with basic passport and descriptor data.*

The major weakness of NPGS database management is not the

functioning of the GRIN, but the paucity of data, even for listed accessions. Few records examined by the committee contained useful information. Evaluation data were particularly lacking. There are bits and pieces of data for various crops, but so much is missing that it is difficult to search the collections of many crops for particular characteristics.

The GRIN staff, of about 10 people, is largely responsible for programming, not data preparation. The DBMU has made major improvements in the network and is developing an increasingly fast and efficient system. However, it is difficult to separate the function of a database management system such as GRIN from the completeness of the database it serves. The NPGS must obtain accurate information (especially passport data, much of which is in PIO records). The accessions not presently listed must be added to the database. The absence of this information makes the network of limited usefulness to most researchers and breeders.

The responsibility for completing GRIN files must fall to the central authority who would control the necessary personnel and funds. Evaluation and characterization descriptors should be reviewed to determine whether or not they are appropriate for the purposes of the network and likely to be completed. For nearly all of the crops examined, a large proportion of descriptors listed have no data. During the course of its investigations, the committee was told of various plans for adding new kinds of data to GRIN. These included data on endangered wild plants, livestock resources, and other genetic resources. Adding these data to GRIN should not be considered until the present NPGS databases are more complete. Increasing and revising network software or enlarging the hardware to allow for the addition of new kinds of data should not, in the committee's view, compete for personnel and funding resources that should first be devoted to completing the present database.

### Accessibility of Data

*The National Plant Germplasm System should continue to seek mechanisms for making the information held in the Germplasm Resources Information Network more easily accessible to scientists and crop specialists in the United States and abroad.*

The network has two basic kinds of clients: those who supply information and those who request it (Mowder and Stoner, 1989). The DBMU has issued about 450 public access codes to persons outside the NPGS. As of February 1990 about 80 of those had logged on to GRIN in the preceding 6 months (J. D. Mowder, U.S. Department of Agriculture, personal communication, January 1990). While many improvements have been made in GRIN that make obtaining information and ordering germplasm easier, data retrieval is slow when compared with the speed of many microcomputers. Many users seeking germplasm do not require a rapid response, and the DBMU has performed searches on request and supplied printed results (Mowder and Stoner, 1989). The NPGS could prepare standardized, printed searches for the more frequently used collections. Individual crop databases should also be made available on diskettes, in a form that can easily be accessed by commercially available software.

## RESEARCH

Research is an essential part of genetic resources management. Improvements in long-term seed preservation, seed viability testing, optimization of regeneration procedures, and determination of population size to minimize drift require mission-oriented research. Advances in tissue culture and cryopreservation may require basic research into the processes underlying cell physiology, development, and regulation. Such initiatives and research activities must complement the principal objectives of the NPGS to preserve, regenerate, document, and distribute plant germplasm.

### Research Agenda

*A research advisory committee should be established to assess and guide the system's research activities.*

While some research effort is undertaken by NPGS scientists, there is no guidance as to the most urgent needs. No group within the national system oversees the development of systemwide research goals and priorities.

NPGS research should improve the acquisition and maintenance of plant germplasm and promote its use. Its goal should not exclusively be to elucidate basic principles, but should include application of technologies and principles to the broad range of NPGS germplasm. Research within the national system should be focused on the problems of germplasm acquisition, maintenance, evaluation, characterization, and use.

The research advisory committee would comment on the research plans of NPGS scientists. Its recommendations could then be used by the NPGRB and the leader of the NPGS in developing budget and programmatic priorities for the national system.

The research advisory committee could also suggest what work might best be accomplished on a contractual or competitive grants basis. This would provide a means for allocating funds for one-time research needs of relatively short duration for which permanent staffing would be inappropriate.

## Program Review

*External peer review of research should be conducted.*

Research programs should be periodically evaluated by a panel of scientists from outside the NPGS. While efforts have been made recently to improve communication between research scientists in the NPGS, the committee felt that the existing in-house peer review of research performance lacked sufficient rigor to ensure that the best possible research is brought to bear on the needs of the national system. Both individual scientific efforts and a site's overall program of research should be reviewed. The reviews should be overseen by the leadership of the NPGS and could be conducted through the proposed research advisory committee.

## Promoting Research

*Funds should be made available for competitive, goal-directed research in areas of specific need.*

There are many capable research scientists who are not part of the national system, but who can undertake research related to plant germplasm. The NPGS must develop a mechanism for providing support to these individuals or institutions to accomplish research essential to its needs. The guidelines for a program of competitive research funding could be developed by the research advisory committee described above.

If funds were provided through the USDA Competitive Research Grants Office, there would be no need to establish a separate program within the NPGS. The research advisory committee's guidelines ensure that only appropriate, mission-oriented, or basic studies would be funded.

Alternatively, the NPGS could administer a limited effort of its own, through its central office and in cooperation with the research advisory committee. One model program was outlined in a report to the National Plant Genetic Resources Board, *Basic Research Support Program for the National Plant Germplasm System* (an unpublished report of the NPGRB approved in the minutes of its October 9–10, 1985, meeting). The goal of this effort was to encourage and facilitate research by scientists in the public and private sectors that would lead to improved germplasm collection, documentation, preservation, maintenance, evaluation, enhancement, and utilization. Scientists in the NPGS should also be allowed to compete for this funding. Research would be directed toward developing new information and methods. The NPGRB report estimated that the program would not cost more than $5 million annually.

Many areas of research are of importance to improving germplasm management. Examples include methods for long-term maintenance of seed and clonal germplasm (e.g., cryopreservation), nondestructive methods for assessing seed viability, elucidation of rapid, reliable techniques for detecting pathogenic organisms in germplasm, and molecular techniques to characterize, evaluate, or identify and enhance germplasm accessions.

## CONCLUSION

Effective management of the nation's germplasm resources is essential to ensure the present and future security of U.S. agriculture. The conclusions of Nelson Klose more than 40 years ago remain true today (Klose, 1950:139).

It seems certain that plant research and introductions of the future not only will contribute new food crops, but will aid as well the progress of mechanical and chemical technology. Often when experimenters develop disease-resistant plant varieties, the disease organisms in turn adjust themselves by developing new virulent strains. Redesigning plants with the desirable characteristics of many species fused into a single new variety offers a limitless challenge to plant workers. Like the introductions of Colonial days, the plants of tomorrow become America's crop heritage for future generations.

As a participant in global efforts to conserve germplasm, the United States can expect greater benefits and responsibilities. The actions recommended in this report are intended to prepare and equip the National Plant Germplasm System as an effective component of national and international agricultural security.

# References

Bramwell, D., O. Hamann, V. Heywood, and H Synge, eds. 1987. Botanic Gardens and the World Conservation Strategy. Orlando, Fla.: Academic Press.

Brown, A. H. D. 1989a. The case for core collections. Pp. 136–156 in The Use of Plant Genetic Resources, A. H. D. Brown, O. H. Frankel, D. R. Marshall, and J. T. Williams, eds. New York: Cambridge University Press.

Brown, A. H. D. 1989b. Core collections: A practical approach to genetic resources management. Genome 31(2):818–824.

Brown, A. H. D., O. H. Frankel, D. R. Marshall, and J. T. Williams, eds. 1989. The Use of Plant Genetic Resources. New York: Cambridge University Press.

Burgess, S., ed. 1971. The National Program for Conservation of Crop Germ Plasm. Athens: University of Georgia.

Chang, T. T. 1985. Principles of genetic conservation. Iowa State Journal of Research 59(4):325–348.

Chang, T. T. 1989. The case for large collections. Pp. 123–135 in The Use of Plant Genetic Resources, A. D. H. Brown, O. H. Frankel, D. R. Marshall, and J. T. Williams, eds. New York: Cambridge University Press.

Chang, T. T., S. M. Dietz, and M. N. Westwood. 1989. Management and utilization of plant germplasm collections. Pp. 127–160 in Biotic Diversity and Germplasm Preservation: Global Imperatives, L. Knutson and A. K. Stoner, eds. Boston: Kluwer Academic Publishers.

Chapman, C. 1987. Criteria for selecting core collections: Basic passport data. International Board for Plant Genetic Resources, Rome, Italy. Photocopy.

Christensen, E. 1989. Sharing the wealth: Methods of compensating source countries for germplasm resources. DIVERSITY 5(1):29–32.

Council for Agricultural Science and Technology. 1984. Plant Germplasm Preservation and Utilization in U.S. Agriculture. Report No. 106, November 1985. Ames, Iowa: Council for Agricultural Science and Technology.

Cox, T. S., J. P. Murphy, and M. M. Goodman. 1988. The contribution of exotic germplasm to American agriculture. Pp. 114–144 in Seeds and Sovereignty: The Use and Control of Plant Genetic Resources, J. R. Kloppenburg, Jr., ed. Durham, N.C.: Duke University Press.

Creech, J. L., and L. P. Reitz. 1971. Plant germ plasm now and for tomorrow. Advances in Agronomy 23:1–49.

Cunningham, I. S. 1984. Frank N. Meyer: Plant Hunter in Asia. Ames: Iowa State University Press.

Dalrymple, D. G. 1986a. Development and Spread of High-Yielding Wheat Varieties in Developing Countries. Washington, D.C.: U.S. Agency for International Development.

Dalrymple, D. G. 1986b. Development and Spread of High-Yielding Rice Varieties in Developing Countries. Washington, D.C.: U.S. Agency for International Development.

Database Management Unit. 1987. The Germplasm Resources Information Network— GRIN: A Brief History and Introduction. Beltsville, Md.: U.S. Department of Agriculture.

Duvick, D. N. 1984. Genetic diversity in major farm crops on the farm and in reserve. Economic Botany 38:161–178.

Ford-Lloyd, B., and M. Jackson. 1986. Plant Genetic Resources: An Introduction to Their Conservation and Use. London: Edward Arnold.

Frankel, O. H., and A. H. D. Brown. 1984. Plant genetic resources today: A critical appraisal. Pp. 249–257 in Crop Genetic Resources: Conservation and Evaluation, J. H. W. Holden and J. T. Williams, eds. London: Allen and Unwin.

General Accounting Office. 1981a. Better Collection and Maintenance Procedures Needed to Help Protect Agriculture's Germplasm Resources. Washington, D.C.: U.S. Government Printing Office.

General Accounting Office. 1981b. The Department of Agriculture Can Minimize the Risk of Potential Crop Failures. Washington, D.C.: U.S. Government Printing Office.

Genetic Resources Conservation Program. 1988. Evaluation of the University of California Tomato Genetics Stock Center: Recommendations for Its Long-Term Management, Funding, and Facilities. Report No. 2. Oakland: University of California Genetic Resources Conservation Program.

Hanson, J., J. T. Williams, and R. Freund. 1984. Institutes Conserving Crop Germplasm: The IBPGR Global Network of Genebanks. Rome, Italy: International Board for Plant Genetic Resources.

Harrison, R. G. 1946. Correspondence, National Academy of Sciences Archives, Washington, D.C., March 1, 1946.

Hodge, W. H., and C. O. Erlanson. 1956. Plant introduction as a federal service. Advances in Agronomy 7:189–211.

Holden, J. H. W. 1984. The second ten years. Pp. 177–185 in Crop Genetic Resources: Conservation and Evaluation, J. H. W. Holden and J. T. Williams, eds. London: George Allen and Unwin.

Hyland, H. 1984. History of plant introduction in the United States. Pp. 5–14 in Plant Genetic Resources: A Conservation Imperative, C. W. Yeatman, D. Kafton, and G. Wilkes, eds. Boulder, Colo.: Westview Press.

International Board for Plant Genetic Resources. 1989. Annual Report 1988. Rome, Italy: International Board for Plant Genetic Resources.

Jones, Q., and S. Gillette. 1982. The NPGS: An Overview 1982. Fort Collins, Colo.: DIVERSITY, Laboratory for Information Science in Agriculture, Colorado State University.

Klose, N. 1950. America's Crop Heritage: The History of Foreign Plant Introduction by the Federal Government. Ames: Iowa State College Press.

McGuire, P. E., and C. O. Qualset, eds. 1990. Genetic Resources at Risk: Scientific Issues,

Technologies, and Funding Policies. Report No. 5. Oakland: University of California Genetic Resources Conservation Program.

Mowder, J. D., and A. K. Stoner. 1989. Plant germplasm information systems. Pp. 419–426 in Biotic Diversity and Germplasm Preservation: Global Imperatives, L. Knutson and A. K. Stoner, eds. Boston: Kluwer Academic Publishers.

Murphy, C. F. 1988. Institutional responsibility of the National Plant Germplasm System. Pp. 204–217 in Seeds and Sovereignty: The Use and Control of Plant Genetic Resources, J. R. Kloppenburg, Jr., ed. Durham, N.C.: Duke University Press.

National Plant Genetic Resources Board. 1984. Plant Germplasm: Conservation and Use. Washington, D.C.: U.S. Department of Agriculture.

National Plant Germplasm Committee. 1986. Operations manual for national clonal germplasm repositories. Agricultural Research Service, U.S. Department of Agriculture, Washington, D.C. Photocopy.

National Research Council. 1972. Genetic Vulnerability of Major Crops. Washington, D.C.: National Academy of Sciences.

National Research Council. 1988. Expansion of the U.S. National Seed Storage Laboratory: Program and Design Considerations. Washington, D.C.: National Academy Press.

Office of Technology Assessment. 1981. An Assessment of the United States Food and Agricultural Research System. OTA-F-155. Washington, D.C.: U.S. Government Printing Office.

Office of Technology Assessment. 1985. Grassroots Conservation of Biological Diversity in the United States. OTA-BP-F-38. Washington, D.C.: U.S. Government Printing Office.

Office of Technology Assessment. 1987. Technologies to Maintain Biological Diversity. OTA-F-330. Washington, D.C.: U.S. Government Printing Office.

Peacock, W. J. 1989. Molecular biology and genetic resources. Pp. 363–376 in The Use of Plant Genetic Resources, A. D. H. Brown, O. H. Frankel, D. R. Marshall, and J. T. Williams, eds. New York: Cambridge University Press.

Peairs, F., L. Brooks, G. Hein, G. Johnson, B. Massey, D. McBride, P. Morrison, J. T. Schultz, and E. Spackman. 1989. Economic Impact of the Russian Wheat Aphid in the Western United States: 1987–1988. Publication No. 129. Fort Collins, Colo.: Great Plains Agricultural Council.

Peeters, J. P., and N. W. Galwey. 1990. Germplasm collections and breeding needs in Europe. Economic Botany 42:503–521.

Plucknett, D. L., and N. J. H. Smith. 1982. Agricultural research and third world food production. Science 217:215–220.

Plucknett, D. L., and N. J. H. Smith. 1986. Sustaining agricultural yields. BioScience 36(1):40–45.

Plucknett, D. L., N. J. H. Smith, J. T. Williams, and N. M. Anishetty. 1987. Gene Banks and the World's Food. Princeton, N.J.: Princeton University Press.

Priestley, D. A. 1986. Seed Aging. Implications for Seed Storage and Persistence in the Soil. Ithaca, N.Y.: Cornell University Press.

Purdue, R. E., Jr., and G. M. Christenson. 1989. Plant exploration. Pp. 67–94 in Plant Breeding Reviews, Vol. 7.: The National Plant Germplasm System of the United States, R. L. Clark, W. W. Roath, and H. L. Shands, eds. Portland, Ore.: Timber Press.

Rick, C. M., J. W. DeVerna, R. T. Chetelat, and M. A. Stevens. 1987. Potential contributions of wide crosses to improvement of processing tomatoes. Acta Horticulturae 200:45–55.

Roos, E. E. 1984a. Genetic shifts in mixed bean populations, part I: Storage effects. Crop Science 24:240–244.

Roos, E. E. 1984b. Genetic shifts in mixed bean populations, part II: Effects of regeneration. Crop Science 24:711–715.

Seed Savers Exchange. 1989a. Garden Seed Inventory, 2d ed. Decorah, Iowa: Seed Saver Publications.

Seed Savers Exchange. 1989b. Fruit, Berry and Nut Inventory, K. Whealy, ed. Decorah, Iowa: Seed Saver Publications.

Shands, H. L., P. J. Fitzgerald, and S. A. Eberhart. 1989. Program for plant germplasm preservation in the United States. Pp. 97–116 in Biotic Diversity and Germplasm Preservation: Global Imperatives, L. Knutson and A. K. Stoner, eds. Boston: Kluwer Academic Publishers.

Shell, E. R. 1990. Seeds in the bank could stave off disaster on the farm. Smithsonian 20(10):94–105.

Simmonds, N. W. 1979. Principles of Crop Improvement. London: Longman.

Stoner, A. K. 1988. Program review, Germplasm Services Laboratory. Plant Sciences Institute, Agricultural Research Service, Beltsville, Maryland. November 14. Photocopy.

Strauss, M. S., J. A. Pino, and J. I. Cohen. 1988. Quantification of diversity in ex-situ plant collections. DIVERSITY 16:30–32.

U.S. Department of Agriculture. 1981. The National Plant Germplasm System: Current Status (1980), Strengths and Weaknesses, Long-Range Plan (1983–1997). Washington, D.C.: U.S. Department of Agriculture.

U.S. Department of Agriculture. 1986. Position Classification—Research Position Evaluation System. Directive 431.3, Personnel Division, Agricultural Research Service, Washington, D.C. September 30.

U.S. Department of Agriculture. 1988a. Plant Inventory No. 196, Part I: Plant Materials Introduced January 1 to June 30, 1987 (Nos 506219 to 510763). Washington, D.C.: U.S. Government Printing Office.

U.S. Department of Agriculture. 1988b. Plant Inventory No. 196, Part II: Plant Materials Introduced July 1 to December 31, 1987 (Nos 510764 to 514275). Washington, D.C.: U.S. Government Printing Office.

U.S. Department of Agriculture. 1989a. Agricultural Statistics 1988. Washington, D.C.: U.S. Government Printing Office.

U.S. Department of Agriculture. 1989b. Economic Indicators of the Farm Sector: National Financial Summary 1988. Washington, D.C.: U.S. Government Printing Office.

White, G. A., and J. A. Briggs. 1989. Plant germplasm acquisition and exchange. Pp. 405–417 in Biotic Diversity and Germplasm Preservation: Global Imperatives, L. Knutson and A. K. Stoner, eds. Boston: Kluwer Academic Publishers.

White, G. A., S. A. Eberhart, P. A. Miller, and J. D. Mowder. 1988. Plant materials registered by Crop Science incorporated into NPGS. Crop Science 28:716–717.

White, G. A., H. L. Shands, and G. R. Lovell. 1989. History and operation of the National Plant Germplasm System. Pp. 5–56 in Plant Breeding Reviews, Vol. 7: The National Plant Germplasm System of the United States, R. L. Clark, W. W. Roath, and H. L. Shands, eds. Portland, Ore.: Timber Press.

Wilkes, H. G. 1988. Plant genetic resources over ten thousand years: From a handful of seed to the crop-specific mega-genebanks. Pp. 67–89 in Seeds and Sovereignty: The Use and Control of Plant Genetic Resources, J. R. Kloppenburg, Jr., ed. Durham, N.C.: Duke University Press.

Witt, S. C. 1985. Brief Book: Biotechnology and Genetic Diversity. San Francisco: California Agricultural Lands Project.

Yazzie, W. B. 1989. Untitled letter. P. 93 in The 1989 Harvest Edition of the Seed Savers Exchange. Decorah, Iowa: Sunrise Publications.

# Glossary

**accession**  A distinct, uniquely identified sample of seeds or plants, that is maintained as part of a germplasm collection.

**active collection**  Comprised of accessions that are maintained under conditions of short- or medium-term storage for the purpose of study, distribution, or use.

**allele**  One of two or more alternative forms of a gene, differing in DNA (deoxyribonucleic acid) nucleotide sequence and affecting the functioning of a single gene product (RNA [ribonucleic acid] or protein). All alleles of a series occupy the same site or locus on each of a pair of homologous chromosomes.

**annual crop**  A crop that is grown from seed to harvest within 1 year.

**base collection**  A comprehensive collection of accessions that are held for the purpose of long-term conservation.

**biological diversity**  The variety and variability among living organisms and the ecological complexes in which they occur.

**breeding line**  A group of plants with similar traits that have been selected for their special combinations of traits from hybrid or other populations. It may be released as a variety or used for further breeding.

**bulking**  The practice of combining several accessions in a collection and managing them as a single accession.

**characterization**  Assessment of the presence, absence, or degree of specific traits that are little influenced in their expression by varying environmental conditions.

**chemotherapy** Treatment with chemicals to eliminate pests or pathogens from a plant or seed sample.

**clonal propagation** The reproduction of plants through asexual means, such as cuttings, grafts, or tissue culture.

**clone** A group of genetically identical individuals that result from asexual, vegetative multiplication; any plant that is propagated vegetatively and that is therefore a genetic duplicate of its parent.

**collection** A sample (e.g., variety, strain, population) maintained at a genetic resources center for the purposes of conservation or use.

**community** A group of ecologically related populations of various species that occur in a particular geographic area at a particular time.

**cryobiology** Study of the effects of extremely low temperatures on biological systems.

**cryopreservation** Maintaining tissues or seeds for the purpose of long-term storage at ultralow temperatures, typically between −150°C and −196°C; produced by storage above or in liquid nitrogen.

**cultivar** A contraction of cultivated variety. *See also* variety.

**cytogenetics** The combined study of cells and genes at the chromosome level.

**electrophoresis** The differential movement of charged molecules in solution through a porous medium in an electric field. The porous medium can be filter paper, cellulose, or, more frequently, a starch or polyacrylamide gel.

**enhancement** The process of improving a germplasm accession by breeding in desirable genes from more agriculturally acceptable cultivars, breeding lines, or other accessions, while retaining the important genetic contributions of the accession. For accessions, such as many of horticultural value, this may entail simple selection following one or more crosses.

**enzyme** A protein produced by living cells that acts as a catalyst in essential chemical reactions in living tissues.

**evaluation** The assessment of plants in a germplasm collection for potentially useful genetic traits, many of which may be environmentally variable (e.g., pest or disease resistance, fruit quality, flavor).

**ex situ conservation** Maintenance or management of an organism away from its native environment. For crop germplasm this term typically refers to maintenance in seed banks or repositories.

**extinct** In the context of this report, the term refers to taxa (e.g., populations, subspecies, species) not found after repeated searches of known and likely areas.

**forage**   Herbaceous plants used as feed for livestock.

**gene**   The basic functional unit of inheritance responsible for the heritability of particular traits.

**genetic diversity**   In a group such as a population or species, the possession of a variety of genetic traits and alleles that frequently result in differing expressions in different individuals.

**genetic resources**   In the context of this report, the term is synonymous with germplasm. *See* germplasm.

**genetic stocks**   Accessions in a collection that typically possess one or more genetic anomalies or aberrations (e.g., multiple or missing chromosomes, unique genetic markers or mutants) that make them of interest for research.

**genome**   A single complete set of the genes or chromosomes of an individual. Typically, gametes such as egg cells contain a single set and are termed haploid, while the somatic cells that comprise the bulk of the living tissue of the plant body contain two sets and are diploid.

**genotype**   In the context of this report, plants with a specific complement of genes.

**germplasm**   Seeds, plants, or plant parts that are useful in crop breeding, research, or conservation. Plants, seed, or cultures that are maintained for the purposes of studying, managing, or using the genetic information they possess.

**grow-out**   The process of growing a plant for the purpose of producing fresh viable seed or for evaluation or characterization.

**heterozygous**   Having one or more unlike alleles at corresponding loci of homologous chromosomes.

**hybrid**   A cross between two different species, races, cultivars, or breeding lines.

**hybridization**   The process of crossing individuals that possess different genetic makeups.

**in situ conservation**   Maintenance or management of an organism within its native environment. For landraces this term includes maintenance in traditional agricultural systems.

**in vitro**   Maintenance or culture of cells, tissues, or plant parts on a sterile, nutrient medium.

**land-grant college**   The Morrill Land-Grant College Act of 1882 provided a trust of public lands, a land grant, to each state to endow a college where practical education in agriculture and engineering could be emphasized. State land-grant colleges and universities were established from this endowment.

**landrace**   A population of plants, typically genetically heterogeneous, commonly developed in traditional agriculture from many years—even centuries—of farmer-directed selection, and which is specifically adapted to local conditions.

**legume**   Any member of the pea family (Leguminosae or alternately, Fabaceae), for example, beans, peanuts, and alfalfa.

**perennial crops**   Crop plants that are managed to be productive over several years. They include herbaceous perennials that die back annually, such as asparagus, and woody perennials with stems that may live for many years, such as apples, citrus crops, or mangos.

**plant genetic resources**   Plants from which the genes needed by breeders and other scientists can be derived. Frequently synonymous with germplasm.

**polymorphic**   In the context of this report, plants with several to many variable forms.

**population**   A group of organisms of the same species that occupy a particular geographic area or region. In general, individuals within a population potentially interbreed with one another.

**quarantine**   For plants, regulatory measures that protect plant species against pests and disease that may be borne on or introduced by imported plants.

**rare**   In the context of this report, the term refers to taxa with small populations that are not currently endangered, but that are at risk of loss.

**regeneration**   Grow-out of a seed accession for the purpose of obtaining a fresh sample with high viability and adequate numbers of seeds.

**restriction fragment length polymorphisms**   Variation that occurs within a species in the length of DNA fragments resulting from digestion of the extracted DNA with one of several enzymes that cleave DNA at specific recognition sites. Changes in the genetic composition result in fragments of altered length.

**seed viability**   The ability of a seed to germinate under appropriate conditions.

**species**   A taxonomic subdivision; a group of organisms that actually or potentially interbreed and are reproductively isolated from other such groups.

**state agricultural experiment station**   Experiment stations were established under the Hatch Act of 1887 that provided annual funding to states to establish agricultural experiment stations under the direction of land-grant colleges.

**thermotherapy**   Treatment of plant materials with heat to eliminate or kill pathogens (e.g., viruses) or pests.

**tissue culture**   A technique for cultivating cells, tissues, or organs of plants in a sterile, synthetic medium; includes the tissues excised from a plant and the culture of pollen or seeds.

**tuber**   A thickened, compressed, fleshy, usually underground stem that may function as a storage organ for food (starch) or water, or for propagation.

**variety**   A plant type within a cultivated species that is distinguishable by one or more characters. When reproduced from seeds or by asexual means (e.g., cuttings) its distinguishing characters are retained. The term is generally considered to be synonymous with cultivar.

**voucher**   A specimen preserved for future reference.

**wild species**   Organisms in or out of captivity that have not been subject to breeding to alter them from their native (wild) state.

# Abbreviations

| | |
|---|---|
| **APHIS** | Animal and Plant Health Inspection Service |
| **ARS** | Agricultural Research Service |
| **CGIAR** | Consultative Group on International Agricultural Research |
| **CIAT** | Centro Internacional de Agricultura Tropical (International Center of Tropical Agriculture) |
| **CPC** | Center for Plant Conservation |
| **CSRS** | Cooperative State Research Service |
| **DBMU** | Database Management Unit |
| **DNA** | deoxyribonucleic acid |
| **FAO** | Food and Agriculture Organization (of the United Nations) |
| **GMT** | Germplasm Matrix Team |
| **GRIN** | Germplasm Resources Information Network |
| **GSL** | Germplasm Services Laboratory |
| **IBPGR** | International Board for Plant Genetic Resources |
| **IR-1** | Interregional Research Project-1 (potatoes) |
| **IR-2** | Interregional Research Project-2 (virus free fruit trees) |
| **Native Seed/ SEARCH** | Native Seed/Southwestern Endangered Aridland Resource Clearing House |
| **NGRL** | National Germplasm Resources Laboratory |
| **NPGC** | National Plant Germplasm Committee |
| **NPGQL** | National Plant Germplasm Quarantine Laboratory |
| **NPGRB** | National Plant Genetic Resources Board |
| **NPGS** | National Plant Germplasm System |

| | |
|---|---|
| **NSF** | National Science Foundation |
| **NSGC** | National Small Grains Collection |
| **NSSL** | National Seed Storage Laboratory |
| **PGOC** | Plant Germplasm Operations Committee |
| **PI** | plant introduction |
| **PIO** | Plant Introduction Office |
| **SCS** | Soil Conservation Service |
| **SSE** | Seed Savers Exchange |
| **TAC** | technical advisory committee |
| **TC** | technical committee |
| **USAID** | U.S. Agency for International Development |
| **USDA** | U.S. Department of Agriculture |

# Authors

**ROBERT W. ALLARD** (*Subcommittee Chairman*)   Allard is emeritus professor of genetics at the University of California, Davis. He has a Ph.D. degree in genetics from the University of Wisconsin. His areas of research include plant population genetics, gene resource conservation, and plant breeding. He is a member of the National Academy of Sciences.

**PAULO DE T. ALVIM**   Since 1963, Alvim has been the scientific director for the Comissão Executiva do Plano da Lavoura Cacaueira, Brasil. He earned a Ph.D. degree from Cornell University with specialization in plant physiology, tropical agriculture, and ecology.

**AMRAM ASHRI**   Since 1971 he has been professor of genetics and breeding at the Hebrew University of Jerusalem, Israel. He has a Ph.D. degree in genetics from the University of California, Davis. His areas of research include plant breeding and the evaluation and utilization of germplasm resources.

**JOHN H. BARTON**   Since 1975 Barton has been a professor of law and director of the International Center on Law and Technology at Stanford University, where he earned his law degree. He is also cofounder of International Technology Management, a consulting firm specializing in international technology, trade, regulation, and transfer. He is a recognized expert on property rights as they relate to genetic resources.

**FREDERICK H. BUTTEL**   Buttel is a professor in the Department of Rural Sociology and a faculty associate in the Program on Science, Technology, and Society at Cornell University. He earned a Ph.D. degree in sociology from the University of Wisconsin-Madison. His areas of interest are in technology and social change, particularly in relation to agricultural research and biotechnology.

**TE-TZU CHANG**   Chang has been head of the International Rice Germplasm Center at the International Rice Research Institute since 1983 and principal scientist since 1985. He has also been a visiting professor at the University of the Philippines, Los Baños, since 1962. He earned a Ph.D. degree in plant genetics and breeding from the University of Minnesota. He had a vital role in the Green Revolution in rice. Chang has broad experience in managing and designing plant genebanks.

**JOHN L. CREECH**   Creech retired from the U.S. Department of Agriculture in 1980. He is presently a senior adviser to the International Board for Plant Genetics Resources in developing genetic resources management systems. His Ph.D. degree from the University of Maryland is in botany. His particular areas of expertise is in plant exploration in the Far East and the Soviet Union.

**PETER R. DAY** (*Committee Chairman*)   Before joining Rutgers University as director of the Center for Agricultural Molecular Biology in 1987, Day was the director of the Plant Breeding Institute, Cambridge, United Kingdom. He has a Ph.D. degree from the University of London, and is a leader in the field of biotechnology and its application to agriculture.

**S. M. (SAM) DIETZ**   From 1966 until his retirement in July 1990, Dietz was coordinator and research leader for the Western Regional Plant Introduction Station, Pullman, Washington. He earned his Ph.D. degree in plant pathology from Washington State University. He has been involved in long-term planning for the National Plant Germplasm System (NPGS), as well as other government studies of the system, and has served as a consultant to overseas agencies.

**ROBERT E. EVENSON**   Since 1977 Evenson has been a professor of economics at Yale University. He has a Ph.D. degree in economics from the University of Chicago. His research interests include agricultural development policy with a special interest in the economics of agricultural research.

**HENRY A. FITZHUGH**  Fitzhugh is deputy director general for research at the International Livestock Center for Africa, Ethiopia. He received a Ph.D. degree in animal breeding from Texas A&M University. His field of research is the development and testing of biological and socioeconomic interventions to improve the productivity of livestock in agricultural production systems.

**MAJOR M. GOODMAN**  Goodman is professor of crop science, statistics, genetics, and botany at North Carolina State University (NCSU) where he has been employed since 1967. He has a Ph.D. degree in genetics from NCSU, and his areas of research are plant breeding, germplasm conservation and utilization, numerical taxonomy, history and evolution of maize, and applied multivariate statistics. Goodman is a member of the National Academy of Sciences.

**JAAP J. HARDON**  In 1985, Hardon became the director of the Center for Genetics Resources, The Netherlands. He has a Ph.D. degree in plant genetics from the University of California. His specialty is plant breeding and genetics.

**VIRGIL A. JOHNSON**  Before his retirement in 1986 Johnson was a research agronomist for the North-Central Region of the Agricultural Research Service, U.S. Department of Agriculture, and is professor emeritus of agronomy at the University of Nebraska, Lincoln. He earned his Ph.D. degree in agronomy from the University of Nebraska. His areas of interest are wheat breeding and genetics, genetics and physiology of wheat, and protein quantity and nutritional quality.

**DONALD R. MARSHALL**  Since 1987 Marshall has been professor of agronomy at the Waite Agricultural Research Institute, University of Adelaide, Australia. He has a Ph.D. degree in genetics from the University of California, Davis. His professional interests are population genetics, plant breeding, host-parasite interactions, and genetic resources conservation.

**A. BRUCE MAUNDER**  Maunder is vice-president of agronomic research at DEKALB Plant Genetics. He has a Ph.D. degree from Purdue University in plant breeding and genetics, and is a well-known sorghum breeder with considerable experience in the use of exotic germplasm.

**CALVIN O. QUALSET** (*Work Group Chairman*)   Qualset is the director of the Genetic Resources Conservation Program (since 1985) and professor of agronomy (since 1967) at the University of California, Davis, where he also earned a Ph.D. degree in genetics. He is active in administration of genetics resources conservation and has special interests in host-plant resistance, quantitative genetics and breeding crops, especially cereals, for agronomic and quality characteristics.

**RAJENDRA S. PARODA**   Paroda is the deputy director general for crop sciences at the Indian Council of Agricultural Research, New Delhi. He has a Ph.D. degree in genetics from the Indian Agricultural Research Institute, New Delhi. He is well known for his contributions as a forage breeder and for his leadership in the field of plant genetic resources in India.

**SETIJATI SASTRAPRADJA**   Sastrapradja is affiliated with the National Center for Research in Biotechnology at the Indonesian Institute of Science. She has a Ph.D. degree in botany from the University of Hawaii.

**CHARLES SMITH**   Smith is a professor of animal breeding strategies at the University of Guelph, Canada. He has a Ph.D. degree in animal breeding from Iowa State University. His research area is in animal breeding strategies, including genetic conservation, and he has been involved in international efforts to conserve domestic animal germplasm.

**JOHN A. SPENCE**   In 1989, Spence was appointed head of the Cocoa Research Unit at the University of the West Indies, Trinidad and Tobago. He has a Ph.D. degree from the University of Bristol, United Kingdom. His research interests are cocoa tissue culture and cryopreservation as alternatives to holding field germplasm collections.

**DAVID H. TIMOTHY**   Timothy has been professor of crop science, botany, and genetics at North Carolina State University since 1966. He has a Ph.D. degree in plant genetics from the University of Minnesota, and has extensive experience in collection, maintenance, characterization, and utilization of germplasm resources of maize and its relatives. His forage breeding research is with subtropical grasses to develop temperate perennial cultivars of superior nutritional quality, yield, and animal performance.

**H. GARRISON WILKES**   Since 1983, Wilkes has been professor of biology at the University of Massachusetts, Boston. He has a Ph.D degree in biology from Harvard University. His field of research is evolution under domestication in cultivated plants, especially maize and its wild relatives, teosinte, and the genus *Tripsacum*.

**LYNDSEY A. WITHERS** Since 1988, Withers has been the in vitro conservation officer in the research program of the International Board for Plant Genetic Resources, Rome, Italy. She has a Ph.D. degree in botany from the University of Nottingham, United Kingdom, and has extensive knowledge of the application of tissue culture, cryopreservation, and plant biotechnology to the conservation of plant genetic resources.

# Index

## A

Administration of NPGS
  advisory structure changes,
    112–114
  barriers to consolidation and
    centralization, 4–5, 88, 107,
    108–112
  budget authority, 89, 107, 108
  crop advisory committees, 8,
    49, 57, 79, 103–104, 114,
    115
  crop curators, 13, 120
  elevation of NPGS within
    ARS, 11, 111–112
  funding for, 9–10, 114
  national collections, 13, 119
  National Plant Genetic Re-
    sources Board, 7–8, 10–12,
    16–17, 102, 109, 112–113
  National Plant Germplasm
    Committee, 8, 12, 102–103,
    112–113
  personnel classification and
    promotion, 97–98, 133–134

  Plant Germplasm Operations
    Committee, 8, 104–105, 114
  reorganization outside ARS,
    3, 10–11, 109–111
  technical committees and
    technical advisory commit-
    tees, 8, 104
  in USDA, 1–2, 87–101; see also
    Agricultural Research Serv-
    ice; Cooperative State
    Research Service
  see also Recommendations
Aegilops species, 5, 45, 58
Agency for International Devel-
    opment, 39, 77, 128, 131
Agricultural experiment sta-
    tions, see State agricultural
    experiment stations
Agricultural Marketing Act of
    1946, 4, 40, 41
Agricultural Research Service
    (ARS)
  administrative role in NPGS,
    2, 7, 9, 43, 47, 49, 52, 57,
    88–98, 108

*161*

# Recent Publications of the Board on Agriculture

*Policy and Resources*

Investing in Research: A Proposal to Strengthen the Agricultural, Food, and Environmental System (1989), 156 pp., ISBN 0-309-04127-9.

Alternative Agriculture (1989), 464 pp., ISBN 0-309-03987-8; ISBN 0-309-03985-1 (pbk).

Understanding Agriculture: New Directions for Education (1988), 80 pp., ISBN 0-309-03936-3.

Designing Foods: Animal Product Options in the Marketplace (1988), 394 pp., ISBN 0-309-03798-0; ISBN 0-309-03795-6 (pbk).

Agricultural Biotechnology: Strategies for National Competitiveness (1987), 224 pp., ISBN 0-309-03745-X.

Regulating Pesticides in Food: The Delaney Paradox (1987), 288 pp., ISBN 0-309-03746-8.

Pesticide Resistance: Strategies and Tactics for Management (1986), 480 pp., ISBN 0-309-03627-5.

Pesticides and Groundwater Quality: Issues and Problems in Four States (1986), 136 pp., ISBN 0-309-03676-3.

Soil Conservation: Assessing the National Resources Inventory, Volume 1 (1986), 134 pp., ISBN 0-309-03649-9.

Soil Conservation: Assessing the National Resources Inventory, Volume 2 (1986), 314 pp., ISBN 0-309-03675-5.

New Directions for Biosciences Research in Agriculture: High-Reward Opportunities (1985), 122 pp., ISBN 0-309-03542-2.

Genetic Engineering of Plants: Agricultural Research Opportunities and Policy Concerns (1984), 96 pp., ISBN 0-309-03434-5.

*Nutrient Requirements of Domestic Animals Series and Related Titles*

Nutrient Requirements of Horses, Fifth Revised Edition (1989), 128 pp., ISBN 0-309-03989-4; diskette included.

Nutrient Requirements of Dairy Cattle, Sixth Revised Edition, Update 1989 (1989), 168 pp., ISBN 0-309-03826-X; diskette included.

Nutrient Requirements of Swine, Ninth Revised Edition (1988), 96 pp., ISBN 0-309-03779-4.

Vitamin Tolerance of Animals (1987), 105 pp., ISBN 0-309-03728-X.

Predicting Feed Intake of Food-Producing Animals (1986), 95 pp., ISBN 0-309-03695-X.

Nutrient Requirements of Cats, Revised Edition (1986), 87 pp., ISBN 0-309-03682-8.

Nutrient Requirements of Dogs, Revised Edition (1985), 79 pp., ISBN 0-309-03496-5.

Nutrient Requirements of Sheep, Sixth Revised Edition (1985), 106 pp., ISBN 0-309-03596-1.

Nutrient Requirements of Beef Cattle, Sixth Revised Edition (1984), 90 pp., ISBN 0-309-03447-7.

Nutrient Requirements of Poultry, Eighth Revised Edition (1984), 71 pp., ISBN 0-309-03486-8.

More information, additional titles (prior to 1984), and prices are available from the National Academy Press, 2101 Constitution Avenue, NW, Washington, DC 20418, (202) 334-3313 (information only); (800) 624-6242 (orders only).